Basic Electrical Construction Drawings

Fenway Park Pavilion Seat Expansion and EMC/State Street Club Project

This expansion illustrates the accomplishment of an ambitious concept in a sensitive historical environment. The $45 million project was completed within the six-month winter off-season and used precise phasing and logistics to ensure complete protection of Fenway Field. The contractor structurally lifted and shored the historic facility to accommodate the installation of a new ring of columns that would ultimately support the significant addition.

26110-11

Trainees with successful module completions may be eligible for credentialing through NCCER's National Registry. To learn more, go to **www.nccer.org** or contact us at **1.888.622.3720.** Our website has information on the latest product releases and training, as well as online versions of our *Cornerstone* newsletter and Pearson's Contren® product catalog.

Your feedback is welcome. You may email your comments to **curriculum@nccer.org,** send general comments and inquiries to **info@nccer.org,** or use the User Update form at the back of this module.

nccer V.1 6/11

Objectives

When you have completed this module, you will be able to do the following:

1. Explain the basic layout of a set of construction drawings.
2. Describe the information included in the title block of a construction drawing.
3. Identify the types of lines used on construction drawings.
4. Using an architect's scale, state the actual dimensions of a given drawing component.
5. Interpret electrical drawings, including site plans, floor plans, and detail drawings.
6. Interpret equipment schedules found on electrical drawings.
7. Describe the type of information included in electrical specifications.

Performance Tasks

Under the supervision of the instructor, you should be able to do the following:

1. Using an architect's scale, state the actual dimensions of a given drawing component.
2. Make a material takeoff of the lighting fixtures specified in Performance Profile Sheet 2 using the drawing provided on Performance Profile Sheet 3. The takeoff requires that all lighting fixtures be counted, and where applicable, the total number of lamps for each fixture type must be calculated.

Trade Terms

Architectural drawings
Block diagram
Blueprint
Detail drawing
Dimensions
Electrical drawing

Elevation drawing
Floor plan
One-line diagram
Plan view
Power-riser diagram
Scale

Schedule
Schematic diagram
Sectional view
Shop drawing
Site plan
Written specifications

Required Trainee Materials

1. Paper and pencil
2. Copy of the latest edition of the *National Electrical Code*®
3. Appropriate personal protective equipment

Note: *NFPA 70*®, *National Electrical Code*®, and *NEC*® are registered trademarks of the National Fire Protection Association, Inc., Quincy, MA 02269. All *National Electrical Code*® and *NEC*® references in this module refer to the 2011 edition of the *National Electrical Code*®.

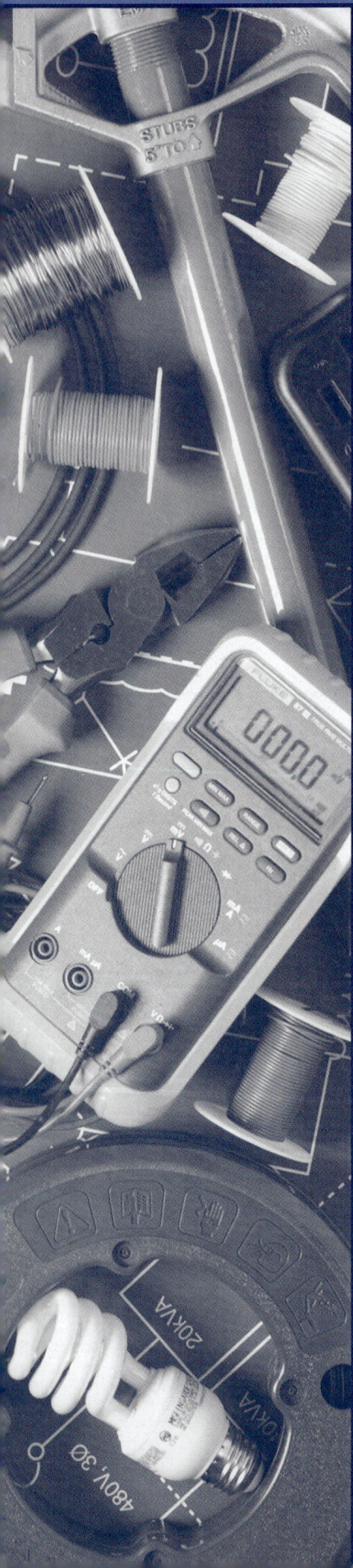

Contents ───────────────────────────────

Topics to be presented in this module include:

Figures and Tables ——————————————————

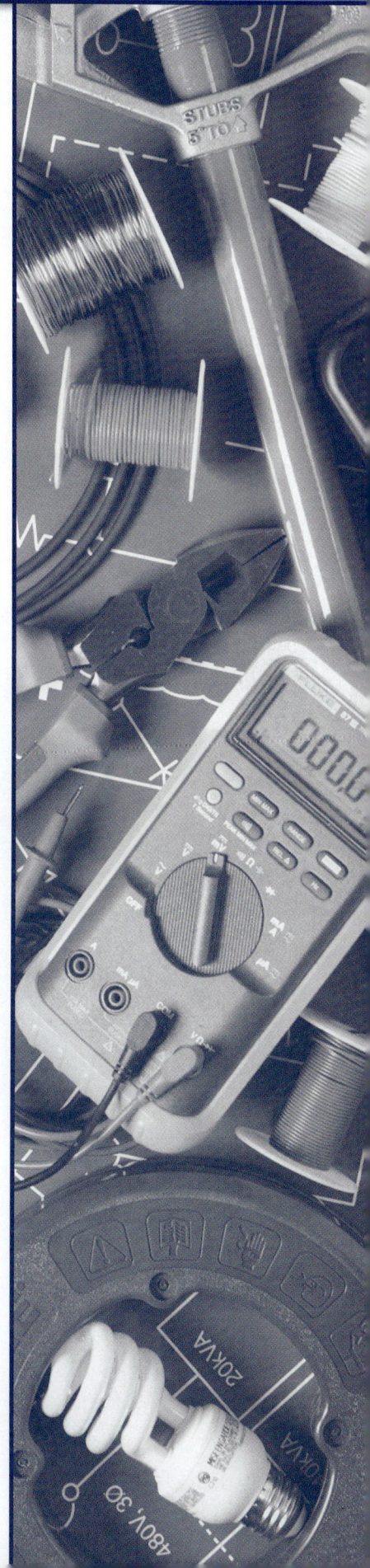

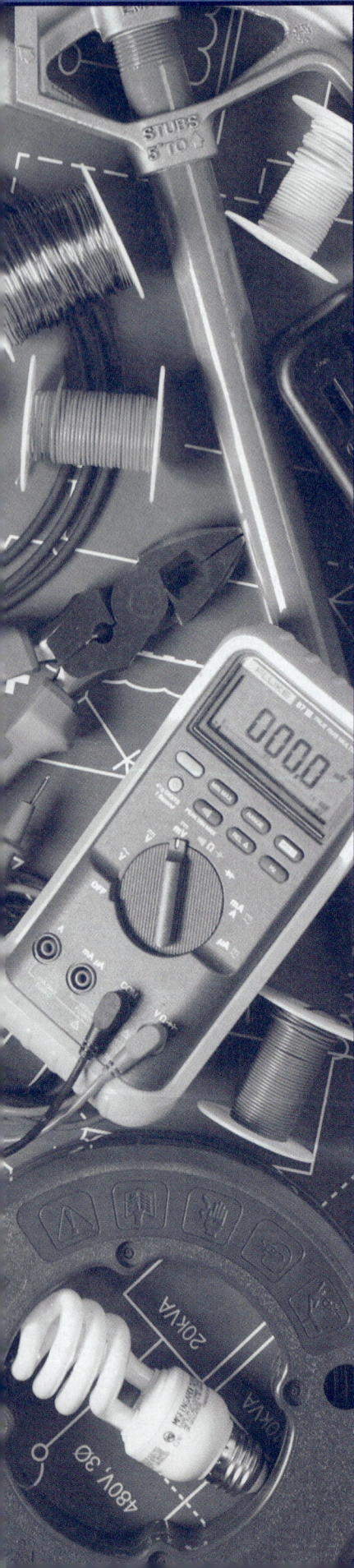

1.0.0 Introduction to Construction Drawings

In all large construction projects and in many of the smaller ones, an architect is commissioned to prepare complete working drawings and specifications for the project. These drawings usually include:

- A site plan indicating the location of the building on the property.
- Floor plans showing the walls and partitions for each floor or level.
- Elevations of all exterior faces of the building.
- Several vertical cross sections to indicate clearly the various floor levels and details of the footings, foundation, walls, floors, ceilings, and roof construction.
- Large-scale detail drawings showing such construction details as may be required.

For projects of any consequence, the architect usually hires consulting engineers to prepare structural, electrical, and mechanical drawings, with the latter encompassing pipefitting, instrumentation, plumbing, and heating, ventilating, and air conditioning drawings.

1.1.0 Site Plan

This type of plan of the building site looks as if the site is viewed from an airplane and shows the property boundaries, the existing contour lines, the new contour lines (after grading), the location of the building on the property, new and existing roadways, all utility lines, and other pertinent details. The drawing scale is also shown. Descriptive notes may also be found on the site (plot) plan listing names of adjacent property owners, the land surveyor, and the date of the survey. A legend or symbol list is also included so that anyone who must work with the site plan can readily read the information. See *Figure 1*.

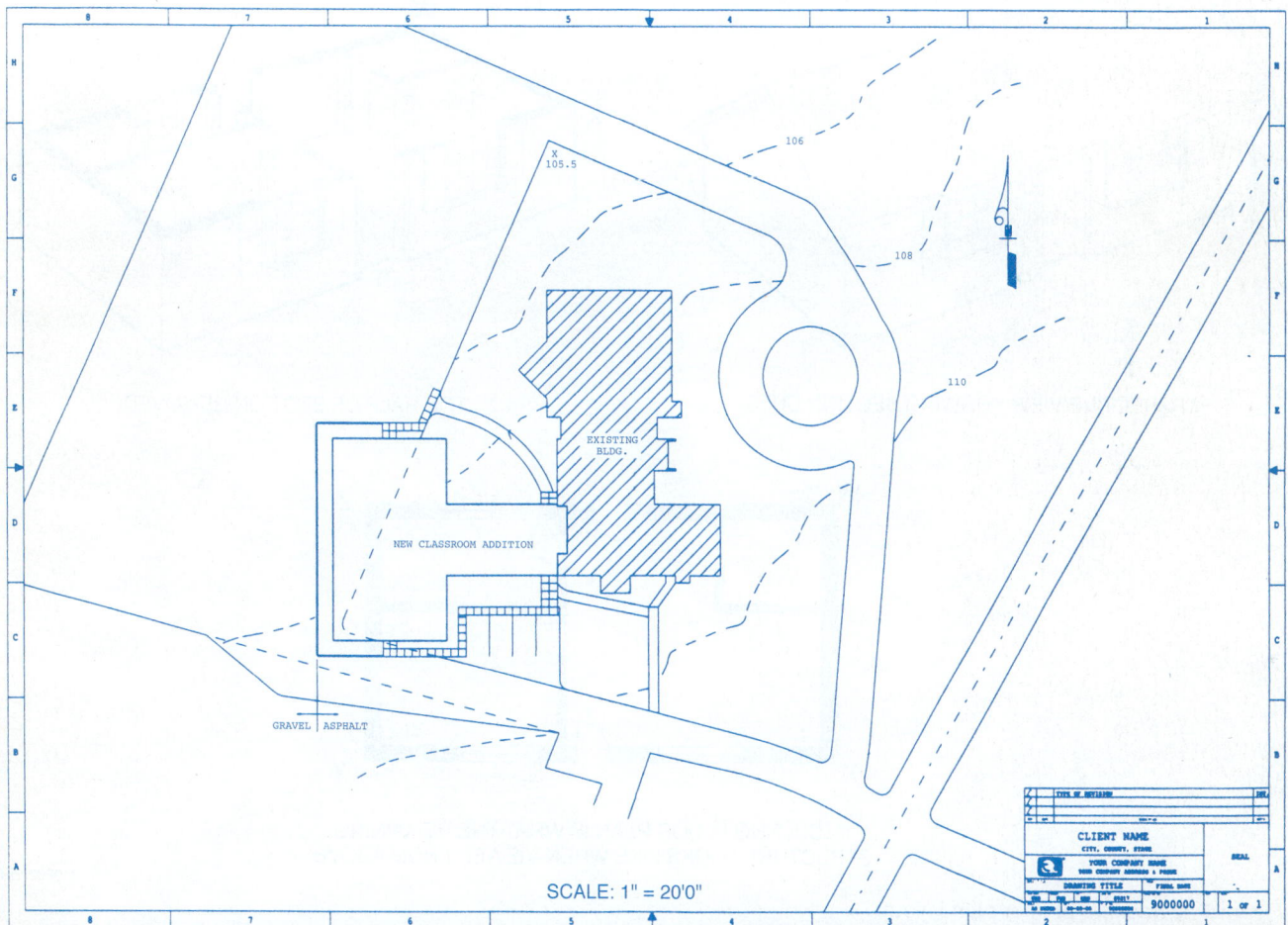

SCALE: 1" = 20'0"

Figure 1 Typical site plan.

110F01.EPS

1.2.0 Floor Plans

The **plan view** of any object is a drawing showing the outline and all details as seen when looking directly down on the object. It shows only two **dimensions,** length and width. The floor plan of a building is drawn as if a horizontal cut were made through the building—at about window height—and then the top portion removed to reveal the bottom part. See *Figure 2*.

If a plan view of a home's basement is needed, the part of the house above the middle of the basement windows is imagined to be cut away. By looking down on the uncovered portion, every detail and partition can be seen. Likewise, imagine the part above the middle of the first floor windows being cut away. A drawing that looks straight down at the remaining part would be called the first floor plan or lower level. A cut through the second floor windows would be called the second floor plan or upper level. See *Figure 3*.

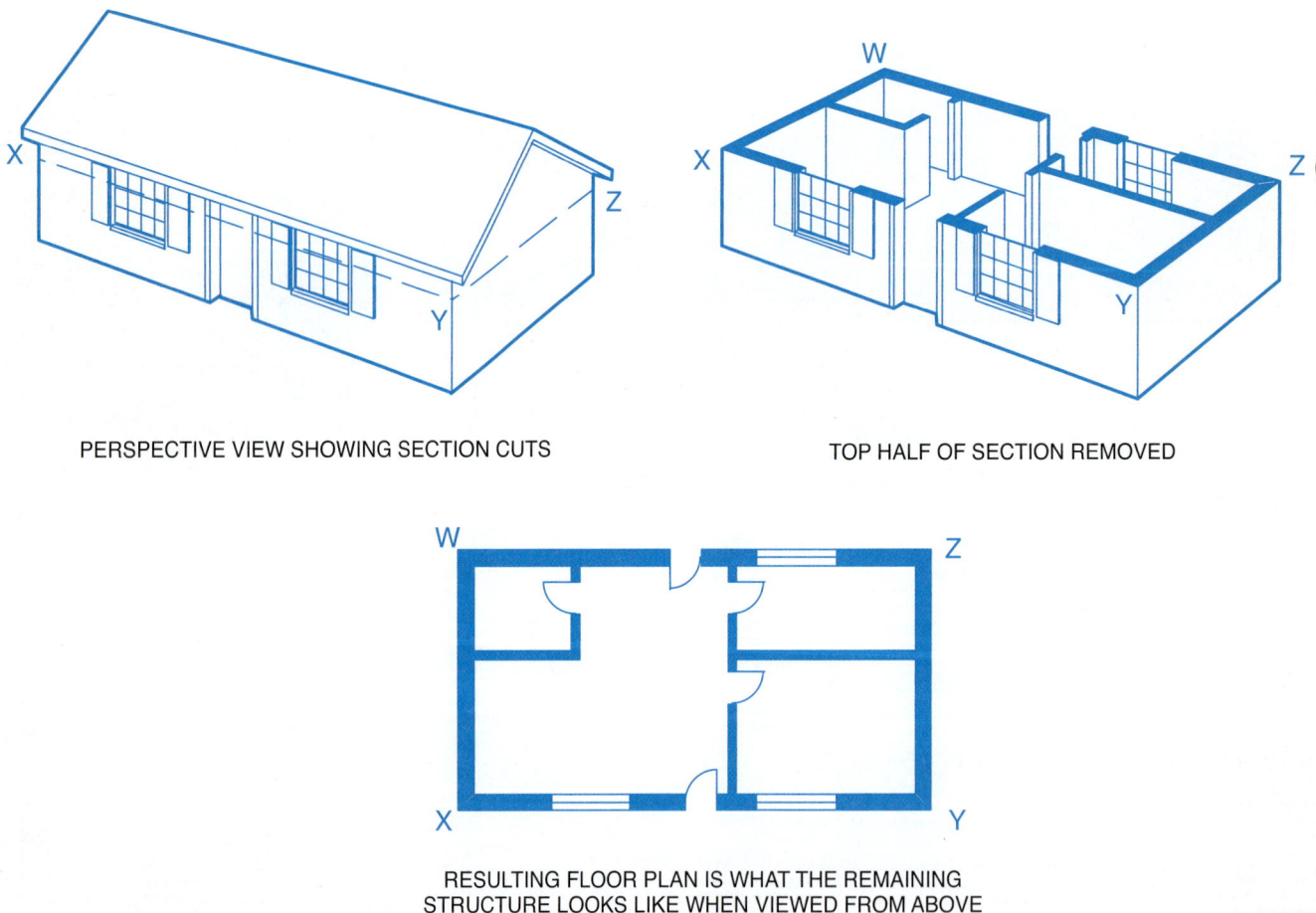

PERSPECTIVE VIEW SHOWING SECTION CUTS

TOP HALF OF SECTION REMOVED

RESULTING FLOOR PLAN IS WHAT THE REMAINING
STRUCTURE LOOKS LIKE WHEN VIEWED FROM ABOVE

110F02.EPS

Figure 2 Principles of floor plan layout.

NCCER — *Contren® Learning Series* 26110-11

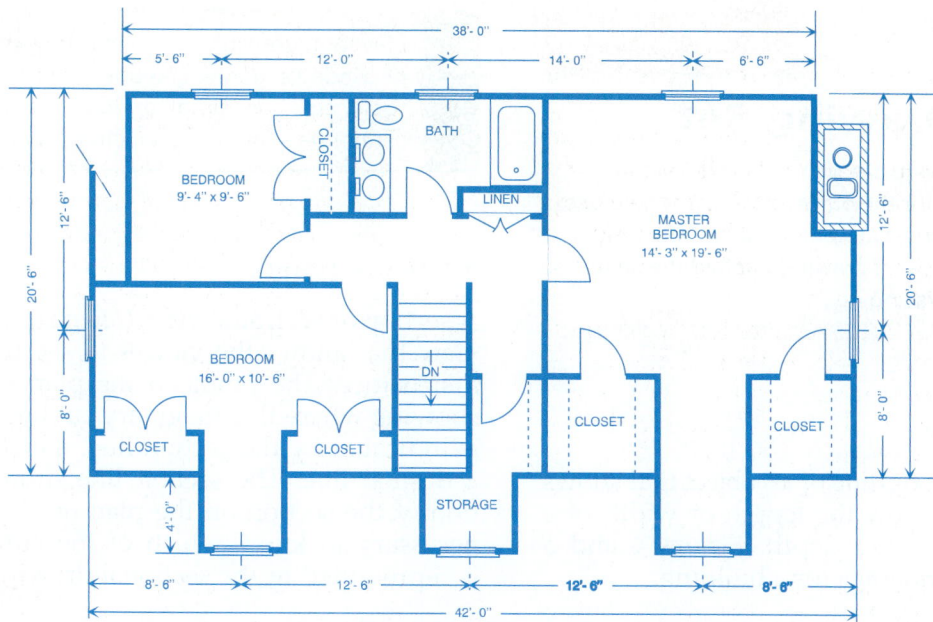

FLOOR PLAN

UPPER LEVEL

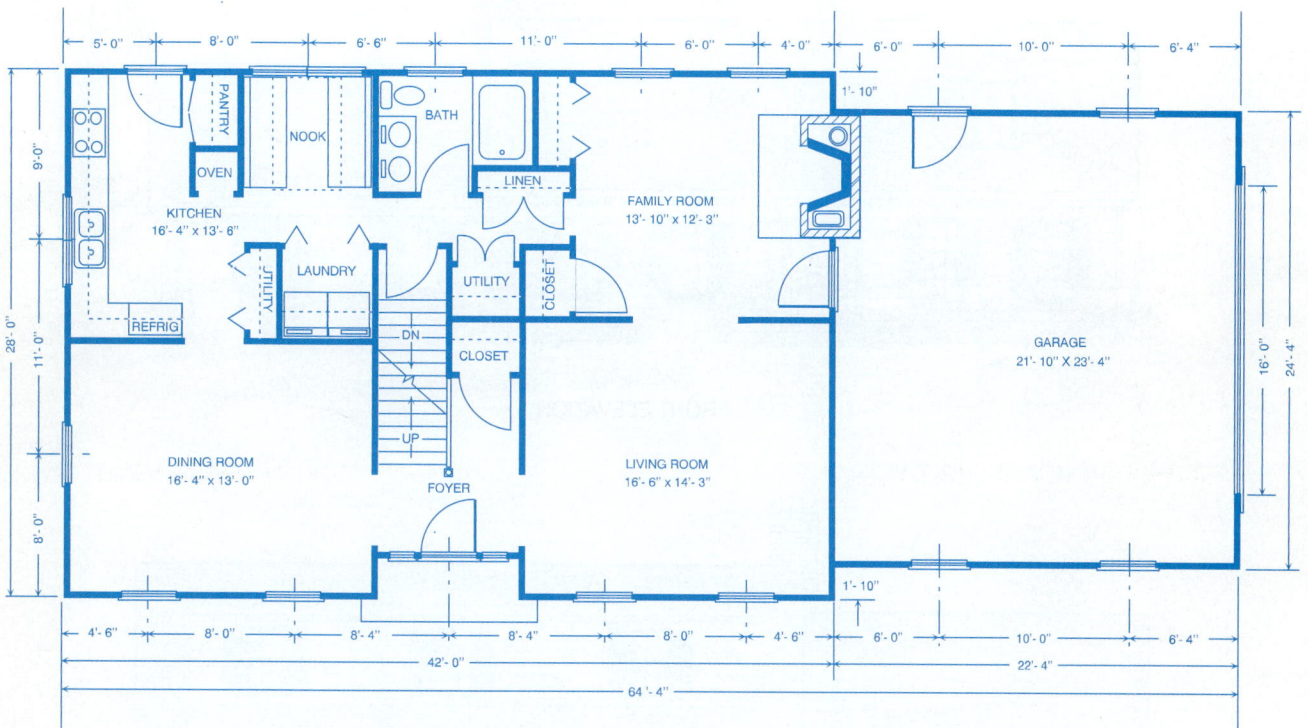

LOWER LEVEL

110F03.EPS

Figure 3 Floor plans of a building.

Using a Drawing Set

Always treat a drawing set with care. It is best to keep two sets, one for the office and one for field use. Be sure to use the most current revision. After you use a sheet from a set of drawings, refold the sheet with the title block facing up.

1.3.0 Elevations

The elevation is an outline of an object that shows heights and may show the length or width of a particular side, but not depth. *Figures 4* and *5* show **elevation drawings** for a building.

1.4.0 Sections

A section or **sectional view** (*Figure 6*) is a cutaway view that allows the viewer to see the inside of a structure. The point on the plan or elevation showing where the imaginary cut has been made is indicated by the section line, which is usually a dashed line. The section line shows the location of the section on the plan or elevation. It is necessary to know which of the cutaway parts is represented in the sectional drawing. To show

FRONT ELEVATION

REAR ELEVATION

110F04.EPS

Figure 4 Front and rear elevations.

LEFT ELEVATION

RIGHT ELEVATION

110F05.EPS

Figure 5 Left and right elevations.

this, arrow points are placed at the ends of the section lines.

In **architectural drawings,** it is often necessary to show more than one section on the same drawing. The different section lines must be distinguished by letters, numbers, or other designations placed at the ends of the lines. These section letters are generally large so as to stand out on the drawings. To further avoid confusion, the same letter is usually placed at each end of the section line. The section is named according to these letters (e.g., Section A-A, Section B-B, and so forth).

A longitudinal section is taken lengthwise while a cross section is usually taken straight across the width of an object. Sometimes, however, a section is not taken along one straight line. It is often taken along a zigzag line to show important parts of the object.

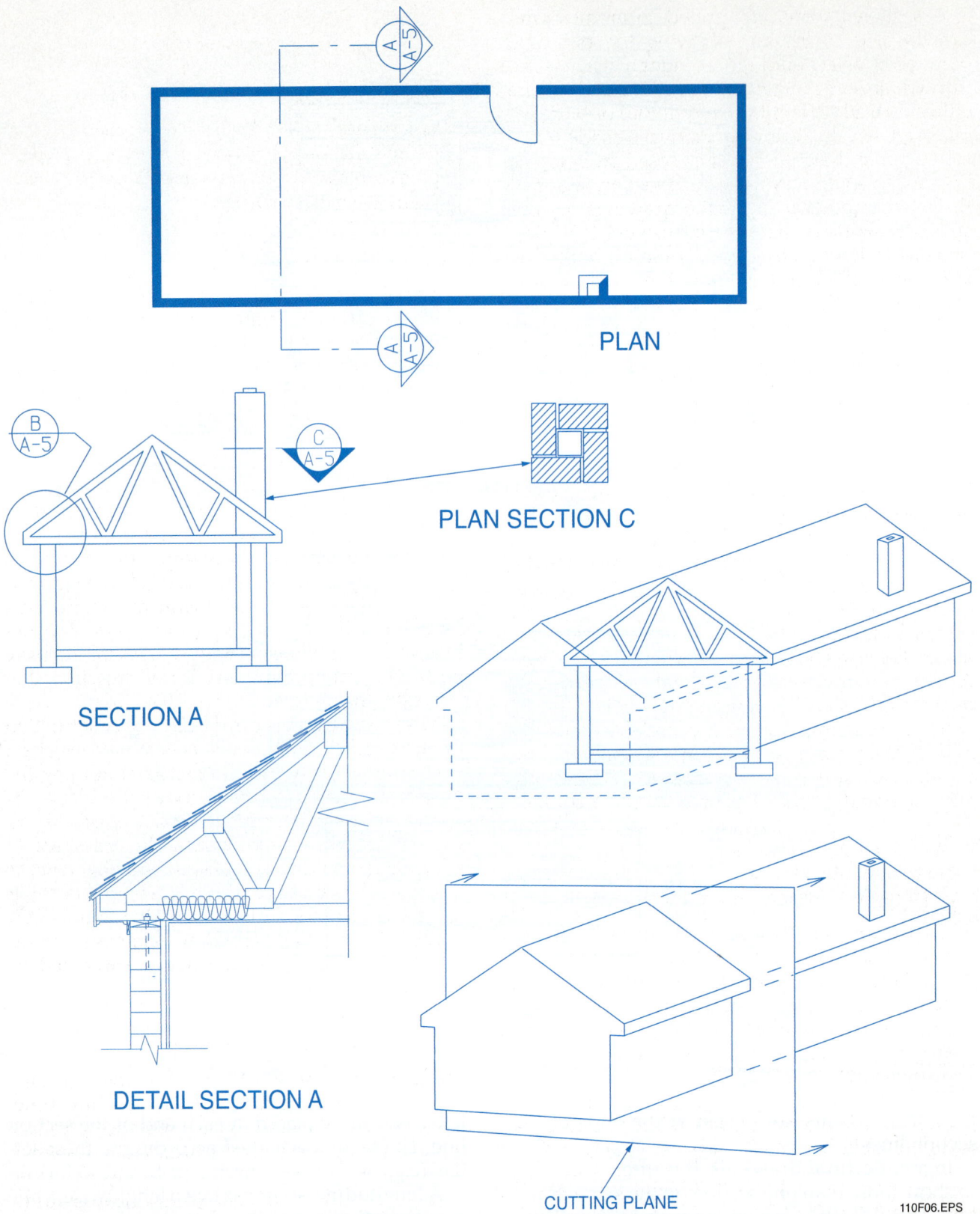

PLAN

PLAN SECTION C

SECTION A

DETAIL SECTION A

CUTTING PLANE

110F06.EPS

Figure 6 Sectional drawing.

A sectional view, as applied to architectural drawings, is a drawing showing the building, or portion of a building, as though it were cut through on some imaginary line. This line may be either vertical (straight up and down) or horizontal. Wall sections are nearly always made vertically so that the cut edge is exposed from top to bottom. In some ways, the wall section is one of the most important of all the drawings to construction workers, because it answers the questions as to how a structure should be built. The floor plans of a building show how each floor is arranged, but the wall sections tell how each part is constructed and usually indicate the material to be used. The electrician needs to know this information when determining wiring methods that comply with the *NEC*®.

1.5.0 Electrical Drawings

Electrical drawings show in a clear, concise manner exactly what is required of the electricians. The amount of data shown on such drawings should be sufficient, but not overdone. This means that a complete set of electrical drawings could consist of only one 8½" × 11" sheet, or it could consist of several dozen 24" × 36" (or larger) sheets, depending on the size and complexity of a given project. A **shop drawing,** for example, may contain details of only one piece of equipment, while a set of working drawings for an industrial installation may contain dozens of drawing sheets detailing the electrical system for lighting and power, along with equipment, motor controls, wiring diagrams, **schematic diagrams,** equipment **schedules,** and a host of other pertinent data.

In general, the electrical working drawings for a given project serve three distinct functions:

- They provide electrical contractors with an exact description of the project so that materials and labor may be estimated to calculate a total cost of the project for bidding purposes.
- They provide workers on the project with instructions as to how the electrical system is to be installed.
- They provide a map of the electrical system once the job is completed to aid in maintenance and troubleshooting for years to come.

Electrical drawings from consulting engineering firms will vary in quality from sketchy, incomplete drawings to neat, precise drawings that are easy to understand. Few, however, will cover every detail of the electrical system. Therefore, a good knowledge of installation practices must go hand-in-hand with interpreting electrical working drawings.

Sometimes electrical contractors will have electrical drafters prepare special supplemental drawings for use by the contractors' employees. On certain projects, these supplemental drawings can save supervision time in the field once the project has begun.

2.0.0 DRAWING LAYOUT

Although a strong effort has been made to standardize drawing practices in the building construction industry, the drawings or **blueprints** prepared by different architectural or engineering firms will rarely be identical. Similarities, however, will exist between most sets of drawings, and with a little experience, you should have no trouble interpreting any set of drawings that might be encountered.

Most drawings used for building construction projects will be drawn on sheets in various sizes. Each drawing sheet has border lines framing the overall drawing and one or more title blocks, as shown in *Figure 7*. The type and size of title blocks varies with each firm preparing the drawings. In addition, some drawing sheets will also contain a revision block near the title block, and perhaps an approval block. This information is normally found on each drawing sheet, regardless of the type of project or the information contained on the sheet.

2.1.0 Title Block

The architect's title block for a drawing is usually boxed in the lower right-hand corner of the drawing sheet; the size of the block varies with the size of the drawing and with the information required. See *Figure 8*.

In general, the title block of an electrical drawing should contain the following information:

- Name of the project
- Address of the project
- Name of the owner or client
- Name of the architectural firm
- Date of completion

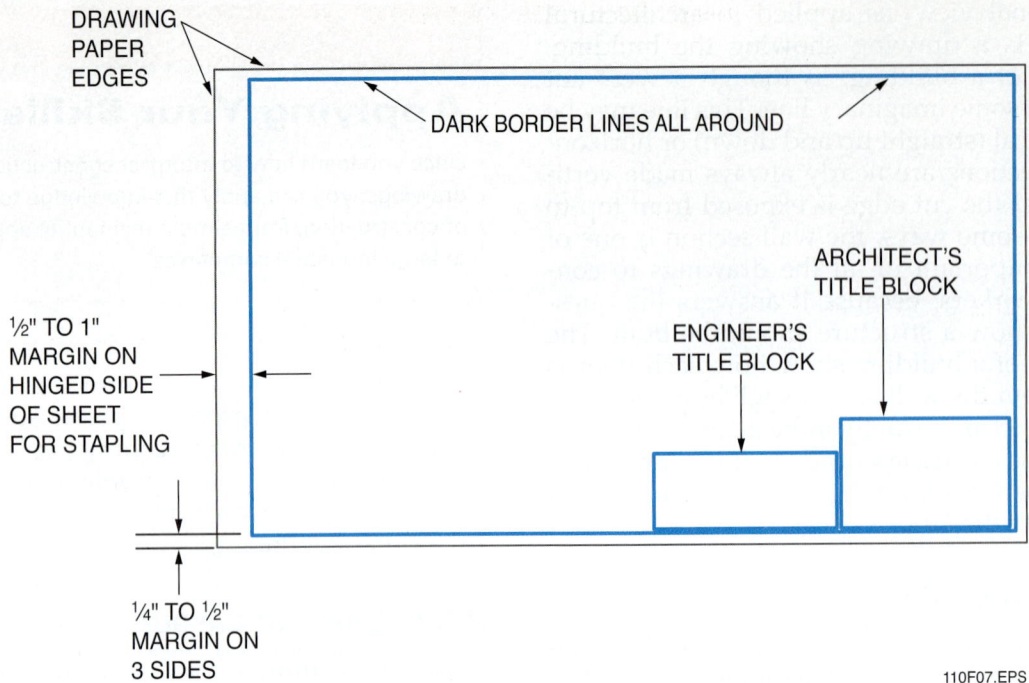

Figure 7 Typical drawing layout.

DRAWING PAPER EDGES

DARK BORDER LINES ALL AROUND

ARCHITECT'S TITLE BLOCK

ENGINEER'S TITLE BLOCK

½" TO 1" MARGIN ON HINGED SIDE OF SHEET FOR STAPLING

¼" TO ½" MARGIN ON 3 SIDES

110F07.EPS

Figure 8 Typical architect's title block.

Professional Stamp

ELECTRICAL

SCALE AS SHOWN

DISTRICT HOME LAUNDRY BUILDING
AUGUSTA COUNTY, VIRGINIA

G. LEWIS CRAIG, ARCHITECT
WAYNESBORO, VIRGINIA

SHEET NO.

E-1

| COMM. NO. | DATE | DRAWN | CHECKED | REVISED |
| 7215 | 9/6/10 | GK | GLC | |

110F08.EPS

Interpreting Electrical Drawings

A good example of when an electrician must interpret the drawings is when wiring a log cabin. The drawings will show the receptacle and switch locations in branch circuits as usual, but the electrician must figure out how to route wires and install boxes where there is no hollow wall and sometimes no ceiling space.

- Scale(s)
- Initials of the drafter, checker, and designer, with dates under each
- Job number
- Sheet number
- General description of the drawing

Often, the consulting engineering firm will also be listed, which means that an additional title block will be applied to the drawing, usually next to the architect's title block. *Figure 9* shows completed architectural and engineering title blocks as they appear on an actual drawing.

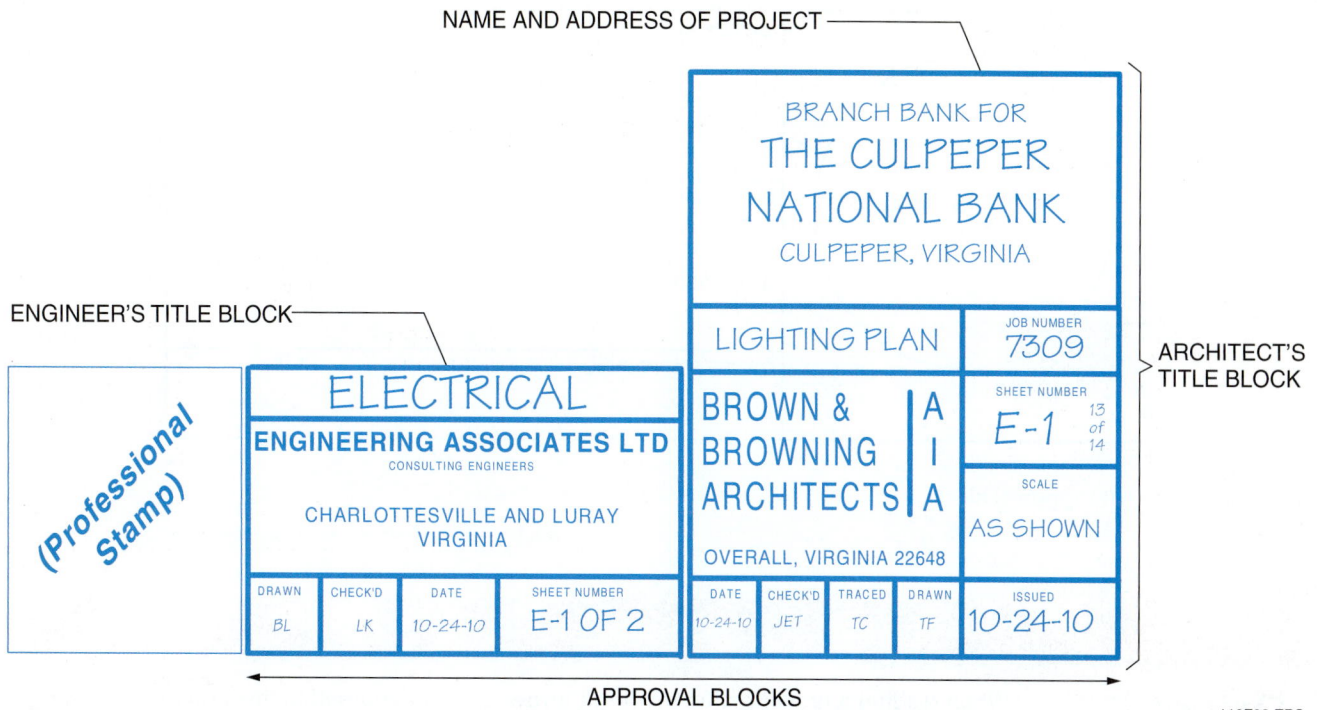

NAME AND ADDRESS OF PROJECT

BRANCH BANK FOR
THE CULPEPER
NATIONAL BANK
CULPEPER, VIRGINIA

ENGINEER'S TITLE BLOCK

(Professional Stamp)

ELECTRICAL
ENGINEERING ASSOCIATES LTD
CONSULTING ENGINEERS

CHARLOTTESVILLE AND LURAY
VIRGINIA

DRAWN	CHECK'D	DATE	SHEET NUMBER
BL	LK	10-24-10	E-1 OF 2

LIGHTING PLAN

JOB NUMBER
7309

BROWN &
BROWNING
ARCHITECTS

A
I
A

OVERALL, VIRGINIA 22648

SHEET NUMBER
E-1 13 of 14

SCALE
AS SHOWN

DATE	CHECK'D	TRACED	DRAWN	ISSUED
10-24-10	JET	TC	TF	10-24-10

ARCHITECT'S TITLE BLOCK

APPROVAL BLOCKS

110F09.EPS

Figure 9 Title blocks.

2.2.0 Approval Block

The approval block, in most cases, will appear on the drawing sheet as shown in *Figure 10*. The various types of approval blocks (drawn, checked, etc.) will be initialed by the appropriate personnel. This type of approval block is usually part of the title block and appears on each drawing sheet.

On some projects, authorized signatures are required before certain systems may be installed, or even before the project begins. An approval block such as the one shown in *Figure 11* indicates that all required personnel have checked the drawings for accuracy, and that the set meets with everyone's approval. Such an approval block usually appears on the front sheet of the blueprint set and may include:

- *Professional stamp* – Registered seal of approval by the licensed architect or consulting engineer.
- *Design supervisor* – Signature of the person who is overseeing the design.
- *Drawn (by)* – Signature or initials of the person who drafted the drawing and the date it was completed.
- *Checked (by)* – Signature or initials of the person who reviewed the drawing and the date of approval.
- *Approved* – Signature or initials of the architect/ engineer and the date of the approval.
- *Owner's approval* – Signature of the project owner or the owner's representative along with the date signed.

COMM. NO.	DATE	DRAWN	CHECKED	REVISED
7215	9/6/10	GK	GLC	

110F10.EPS

Figure 10 Typical approval block.

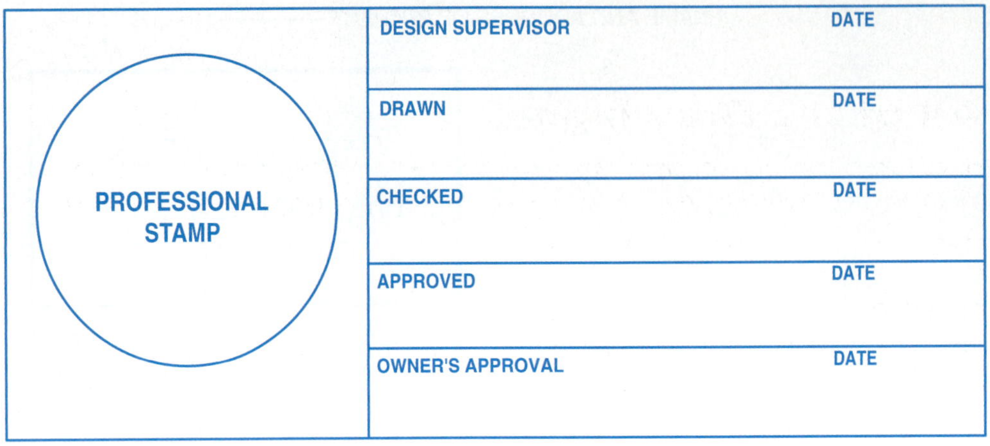

	DESIGN SUPERVISOR	DATE
	DRAWN	DATE
PROFESSIONAL STAMP	CHECKED	DATE
	APPROVED	DATE
	OWNER'S APPROVAL	DATE

110F11.EPS

Figure 11   Alternate approval block.

Orient Yourself

When reading a drawing, find the north arrow to orient yourself to the structure. Knowing where north is enables you to accurately describe the locations of walls and other parts of the building.

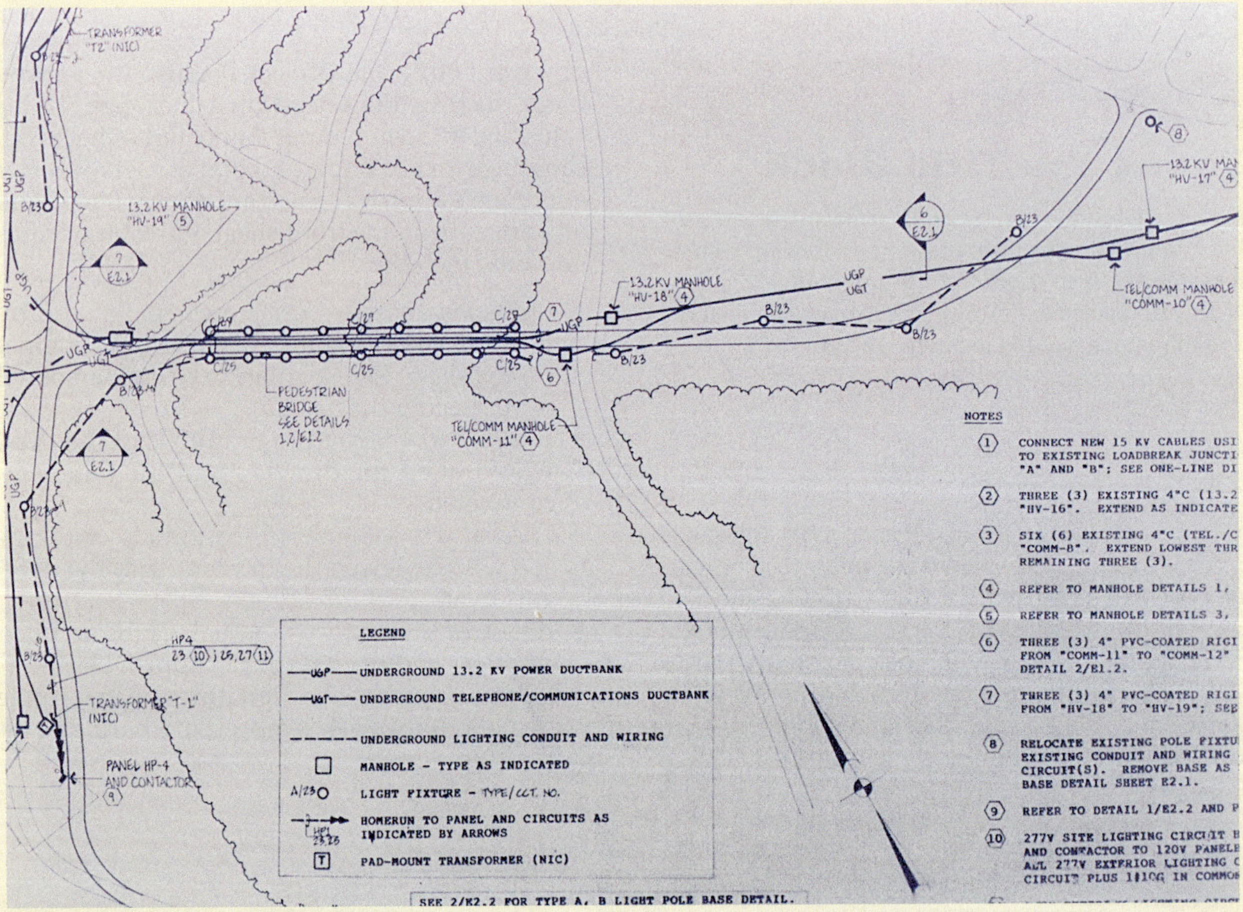

110SA01.EPS

NCCER — *Contren® Learning Series* 26110-11

Using All of the Drawings

Look back over the information on floor plans, elevations, and sections. What kinds of information would an electrician get from each of these drawings? What could a sectional drawing show that a floor plan could not?

2.3.0 Revision Block

Sometimes electrical drawings will have to be partially redrawn or modified during the construction of a project. It is extremely important that such modifications are noted and dated on the drawings to ensure that the workers have an up-to-date set of drawings to work from. In some situations, sufficient space is left near the title block for dates and descriptions of revisions, as shown in *Figure 12*. In other cases, a revision block is provided (again, near the title block), as shown in *Figure 13*. The area on the drawing where the revision has been made will often be circled with a cloud shape.

REVISIONS
10/12/10 - REVISED LIGHTING FIXTURE
NO. 3 IN. LIGHTING FIXTURE SCHEDULE

ELECTRICAL
DISTRICT HOME
LAUNDRY BUILDING
AUGUSTA COUNTY, VIRGINIA

G. LEWIS CRAIG, ARCHITECT
WAYNESBORO, VIRGINIA

Professional Stamp

COMM. NO.	DATE	DRAWN	CHECKED	REVISED	SHEET NO.
7215	9/6/10	GK	GLC	TF	E-1

110F12.EPS

Figure 12   One method of showing revisions on working drawings.

REVISIONS					
REV	DESCRIPTION	DR	APPD	DATE	
1	FIXTURE NO. 3 IN. LIGHTING-FIXTURE SCHDL	GK	GLC	10/12/10	

ELECTRICAL
DISTRICT HOME
LAUNDRY BUILDING
AUGUSTA COUNTY, VIRGINIA

G. LEWIS CRAIG, ARCHITECT
WAYNESBORO, VIRGINIA

Professional Stamp

COMM. NO.	DATE	DRAWN	CHECKED	REVISED	SHEET NO.
7215	9/6/10	GK	GLC	TF	E-1

110F13.EPS

Figure 13   Alternative method of showing revisions on working drawings.

NOTE

Architects, engineers, designers, and drafters have their own methods of showing revisions, so expect to find deviations from those shown here.

CAUTION

When a set of electrical working drawings has been revised, always make certain that the most up-to-date set is used for all future layout work. Either destroy the old, obsolete set of drawings or else clearly mark on the affected sheets, *Obsolete Drawing—Do Not Use.* Also, when working with a set of working drawings and written specifications for the first time, thoroughly check each page to see if any revisions or modifications have been made to the originals. Doing so can save much time and expense to all concerned with the project.

3.0.0 DRAFTING LINES

You will encounter many types of drafting lines. To specify the meaning of each type of line, contrasting lines can be made by varying the width of the lines or breaking the lines in a uniform way.

Figure 14 shows common lines used on architectural drawings. However, these lines can vary. Architects and engineers have strived for a common standard for the past century, but unfortunately, their goal has yet to be reached. Therefore, you will find variations in lines and symbols from drawing to drawing, so always consult the legend or symbol list when referring to any drawing. Also, carefully inspect each drawing to ensure that line types are used consistently.

The drafting lines shown in *Figure 14* are used as follows:

- *Light full line* – This line is used for section lines, building background (outlines), and similar

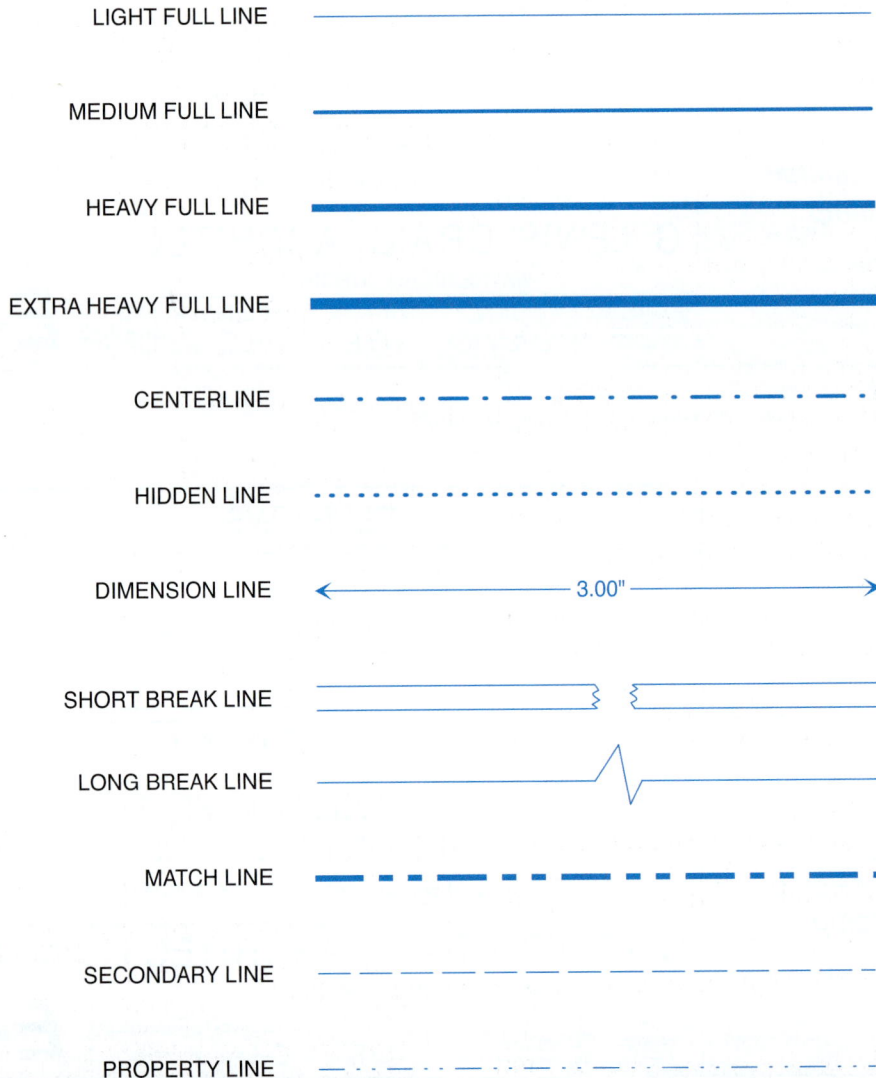

LIGHT FULL LINE

MEDIUM FULL LINE

HEAVY FULL LINE

EXTRA HEAVY FULL LINE

CENTERLINE

HIDDEN LINE

DIMENSION LINE ← 3.00" →

SHORT BREAK LINE

LONG BREAK LINE

MATCH LINE

SECONDARY LINE

PROPERTY LINE

110F14.EPS

Figure 14   Typical drafting lines.

uses where the object to be drawn is secondary to the system being shown (e.g., HVAC or electrical).

- *Medium full line* – This type of line is frequently used for hand lettering on drawings. It is further used for some drawing symbols, circuit lines, etc.
- *Heavy full line* – This line is used for borders around title blocks, schedules, and for hand lettering drawing titles. Some types of symbols are frequently drawn with a heavy full line.
- *Extra heavy full line* – This line is used for border lines on architectural/engineering drawings.
- *Centerline* – A centerline is a broken line made up of alternately spaced long and short dashes. It indicates the centers of objects such as holes, pillars, or fixtures. Sometimes, the centerline indicates the dimensions of a finished floor.
- *Hidden line* – A hidden line consists of a series of short dashes that are closely and evenly spaced. It shows the edges of objects that are not visible in a particular view. The object outlined by hidden lines in one drawing is often fully pictured in another drawing.
- *Dimension line* – These are thin lines used to show the extent and direction of dimensions. The dimension is usually placed in a break inside the dimension lines. Normal practice is to place the dimension lines outside the object's outline. However, it may sometimes be necessary to draw the dimensions inside the outline.
- *Short break line* – This line is usually drawn freehand and is used for short breaks.
- *Long break line* – This line, which is drawn partly with a straightedge and partly with freehand zigzags, is used for long breaks.
- *Match line* – This line is used to show the position of the cutting plane. Therefore, it is also called the cutting plane line. A match or cutting plane line is a heavy line with long dashes alternating with two short dashes. It is used on drawings of large structures to show where one drawing stops and the next drawing starts.
- *Secondary line* – This line is frequently used to outline pieces of equipment or to indicate reference points of a drawing that are secondary to the drawing's purpose.
- *Property line* – This is a light line made up of one long and two short dashes that are alternately spaced. It indicates land boundaries on the site plan.

Other uses of the lines just mentioned include the following:

- *Extension lines* – Extension lines are lightweight lines that start about $\frac{1}{16}$ inch away from the edge of an object and extend out. A common use of extension lines is to create a boundary for dimension lines. Dimension lines meet extension lines with arrowheads, slashes, or dots. Extension lines that point from a note or other reference to a particular feature on a drawing are called leaders. They usually end in either an arrowhead or a dot and may include an explanatory note at the end.
- *Section lines* – These are often referred to as cross-hatch lines. Drawn at a 45° angle, these lines show where an object has been cut away to reveal the inside.
- *Phantom lines* – Phantom lines are solid, light lines that show where an object will be installed. A future door opening or a future piece of equipment can be shown with phantom lines.

3.1.0 Electrical Drafting Lines

Besides the architectural lines shown in *Figure 14* consulting electrical engineers, designers, and drafters use additional lines to represent circuits and their related components. Again, these lines may vary from drawing to drawing, so check the symbol list or legend for the exact meaning of lines on the drawing with which you are working. *Figure 15* shows lines used on some electrical drawings.

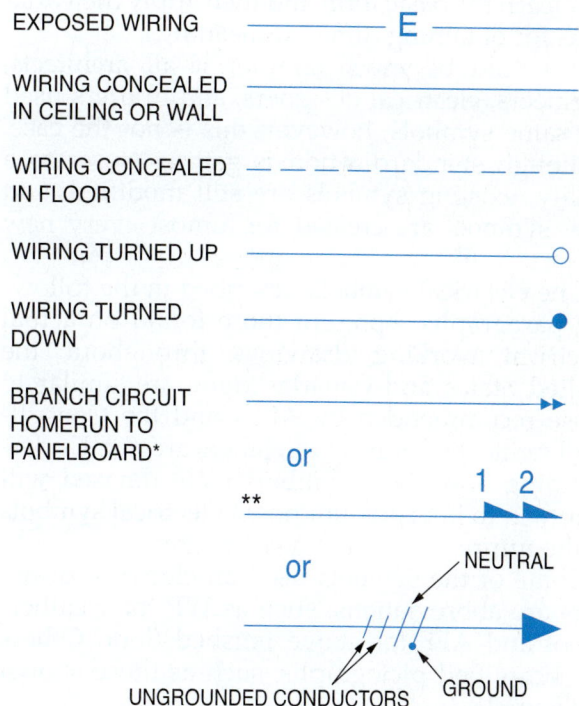

* Number of arrowheads indicates number of circuits. A number at each arrowhead may be used to identify circuit numbers.

** Half arrowheads are sometimes used for homeruns to avoid confusing them with drawing callouts.

110F15.EPS

Figure 15   Electrical drafting lines.

4.0.0 ELECTRICAL SYMBOLS

The electrician must be able to correctly read and understand electrical working drawings. This includes a thorough knowledge of electrical symbols and their applications.

An electrical symbol is a figure or mark that stands for a component used in the electrical system. *Figure 16* shows a list of electrical symbols that are currently recommended by the American National Standards Institute (ANSI). It is evident from this list of symbols that many have the same basic form, but, because of some slight difference, their meaning changes. For example, the receptacle symbols in *Figure 17* each have the same basic form (a circle), but the addition of a line or an abbreviation gives each an individual meaning. A good procedure to follow in learning symbols is to first learn the basic form and then apply the variations for obtaining different meanings.

It would be much simpler if all architects, engineers, electrical designers, and drafters used the same symbols; however, this is not the case. Although standardization is getting closer to a reality, existing symbols are still modified, and new symbols are created for almost every new project.

The electrical symbols described in the following paragraphs represent those found on actual electrical working drawings throughout the United States and Canada. Many are similar to those recommended by ANSI and the Consulting Engineers Council/US; others are not. Understanding how these symbols were devised will help you to interpret unknown electrical symbols in the future.

Some of the symbols used on electrical drawings are abbreviations, such as WP for weatherproof and AFF for above finished floor. Others are simplified pictographs, such as those shown in *Figure 18*.

In some cases, the symbols are combinations of abbreviations and pictographs, such as in *Figure 18* for a fusible safety switch, a nonfusible safety switch, and a double-throw safety switch. In each example, a pictograph of a switch enclosure has been combined with an abbreviation:

F (fusible), DT (double-throw), and NF (nonfusible), respectively.

Lighting outlet symbols have been devised that represent incandescent, fluorescent, and high-intensity discharge lighting; a circle usually represents an incandescent fixture, and a rectangle is used to represent a fluorescent fixture. These symbols are designed to indicate the physical shape of a particular fixture, and while the circles representing incandescent lamps are frequently enlarged somewhat, symbols for fluorescent fixtures are usually drawn as close to scale as possible. The type of mounting used for all lighting fixtures is usually indicated in a lighting fixture schedule, which is shown on the drawings or in the written specifications.

The type of lighting fixture is identified by a numeral placed inside a triangle or other symbol, and placed near the fixture to be identified. A complete description of the fixtures identified by the symbols must be given in the lighting fixture schedule and should include the manufacturer, catalog number, number and type of lamps, voltage, finish, mounting, and any other information needed for proper installation of the fixture.

Switches used to control lighting fixtures are also indicated by symbols (usually the letter S followed by numerals or letters to define the exact type of switch). For example, S_3 indicates a three-way switch; S_4 identifies a four-way switch; and S_p indicates a single-pole switch with a pilot light. A subscript letter is often used to identify the fixtures that are controlled by that switch.

Main distribution centers, panelboards, transformers, safety switches, and other similar electrical components are indicated by electrical symbols on floor plans and by a combination of symbols and semipictorial drawings in riser diagrams.

A detailed description of the service equipment is usually given in the panelboard schedule or in the written specifications. However, on small projects, the service equipment is sometimes indicated only by notes on the drawings.

Circuit and feeder wiring symbols are getting closer to being standardized. Most circuits concealed in the ceiling or wall are indicated by a solid line; a broken line is used for circuits concealed in the floor or ceiling below; and exposed raceways are indicated by short dashes or else the letter *E* placed in the same plane with the circuit line at various intervals. The number of conductors in a conduit or raceway system may be indicated in the panelboard schedule under the appropriate column, or the information may be shown on the floor plan.

Symbols for communication and signal systems, as well as symbols for light and power,

SWITCH OUTLETS

Single-Pole Switch — S

Double-Pole Switch — S$_2$

Three-Way Switch — S$_3$

Four-Way Switch — S$_4$

Key-Operated Switch — S$_K$

Switch w/Pilot — S$_P$

Low-Voltage Switch — S$_L$

Switch & Single Receptacle — ⊖$_S$

Switch & Duplex Receptacle — ⊖$_S$

Door Switch — S$_D$

Momentary Contact Switch — S$_{MC}$

RECEPTACLE OUTLETS

Single Receptacle

Duplex Receptacle

Triplex Receptacle

Split-Wired Duplex Recep.

Single Special Purpose Recep.

Duplex Special Purpose Recep.

Range Receptacle — R

Special Purpose Connection or Provision for Connection. Sub-script letters indicate Function (DW - Dishwasher; CD - Clothes Dryer, etc.) — DW

Clock Receptacle w/Hanger — C

Fan Receptacle w/Hanger — F

Single Floor Receptacle

Note: A numeral or letter within the symbol or as a subscript keyed to the list of symbols indicates type of receptacle or usage.

LIGHTING OUTLETS

	Ceiling	Wall
Surface Fixture	○	─○
Surface Fixt. w/Pull Chain	○PC	─○PC
Recessed Fixture	Ⓡ	─Ⓡ
Surface or Pendant Fluorescent Fixture	[○]	
Recessed Fluor. Fixture	[○R]	
Surface or Pendant Continuous Row Fluor. Fixtures	[○]	
Recessed Continuous Row Fluorescent Fixtures	[○R]	
Surface Exit Light	Ⓧ	─Ⓧ
Recessed Exit Light	ⓍⓇ	─ⓍⓇ
Blanked Outlet	Ⓑ	─Ⓑ
Junction Box	Ⓙ	─Ⓙ

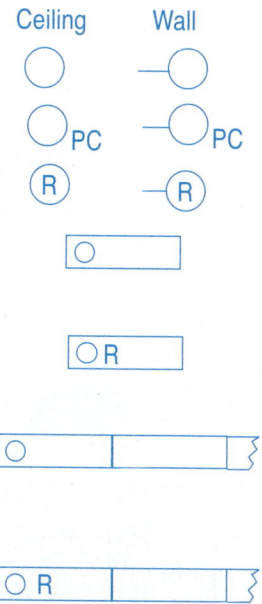

CIRCUITING

Wiring Concealed in Ceiling or Wall — ————

Wiring Concealed in Floor — – – – –

Wiring Exposed — ············

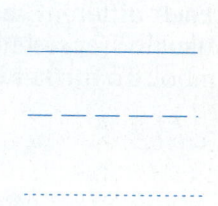

Branch Circuit Homerun to Panelboard. Number of arrows indicates number of circuits in run. Note: Any circuit without further identification is 2-wire. A greater number of wires is indicated by cross lines as shown below. Wire size is sometimes shown with numerals placed above or below cross lines. — ◄◄─

———/ / /——— 3-Wire

———/ / / /——— 4-Wire

Figure 16 ANSI electrical symbols.

110F16.EPS

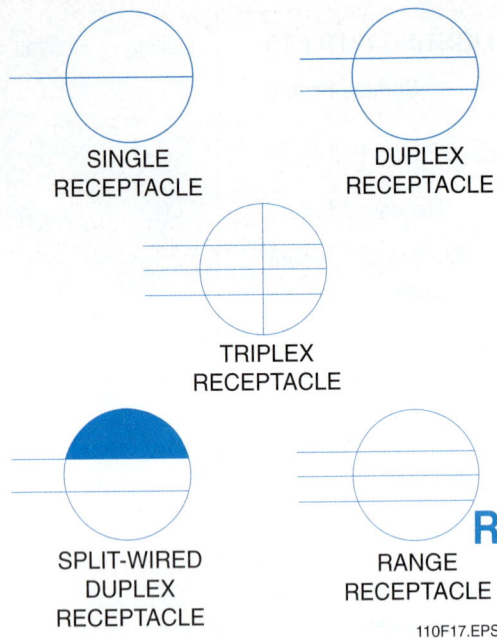

Figure 17 Various receptacle symbols used on electrical drawings.

SINGLE RECEPTACLE

DUPLEX RECEPTACLE

TRIPLEX RECEPTACLE

SPLIT-WIRED DUPLEX RECEPTACLE

RANGE RECEPTACLE

110F17.EPS

are drawn to an appropriate scale and accurately located with respect to the building. This reduces the number of references made to the architectural drawings. Where extreme accuracy is required in locating outlets and equipment, exact dimensions are given on larger-scale drawings and shown on the plans.

Each different category in an electrical system is usually represented by a basic distinguishing symbol. To further identify items of equipment or outlets in the category, a numeral or other identifying mark is placed within the open basic symbol. In addition, all such individual symbols used on the drawings should be included in the symbol list or legend. The electrical symbols shown in *Figure 19* were modified by a consulting engineering firm for use on a small industrial electrical installation. The symbols shown in *Figure 20* are those recommended by the Consulting Engineers Council/US. You should become familiar with these symbols.

5.0.0 SCALE DRAWINGS

In most electrical drawings, the components are so large that it would be impossible to draw them actual size. Consequently, drawings are made to some reduced scale; that is, all the distances are drawn smaller than the actual dimensions of the object itself, with all dimensions being reduced in the same proportion. For example, if a floor plan of a building is to be drawn to a scale of ¼" = 1'–0", each ¼" on the drawing would equal 1 foot on the building itself; if the scale is ⅛" = 1'–0", each ⅛" on the drawing equals 1 foot on the building, and so forth.

When architectural and engineering drawings are produced, the selected scale is very important. Where dimensions must be held to extreme accuracy, the scale drawings should be made as large as practical with dimension lines added. Where dimensions require only reasonable accuracy, the object may be drawn to a smaller scale (with dimension lines possibly omitted).

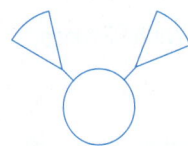

DOUBLE FLOODLIGHT FIXTURE

INFRARED ELECTRIC HEATER WITH TWO QUARTZ LAMPS

F — FUSIBLE SAFETY SWITCH

NF — NON-FUSIBLE SAFETY SWITCH

DT — DOUBLE-THROW SAFETY SWITCH

110F18.EPS

Figure 18 General types of symbols used on electrical drawings.

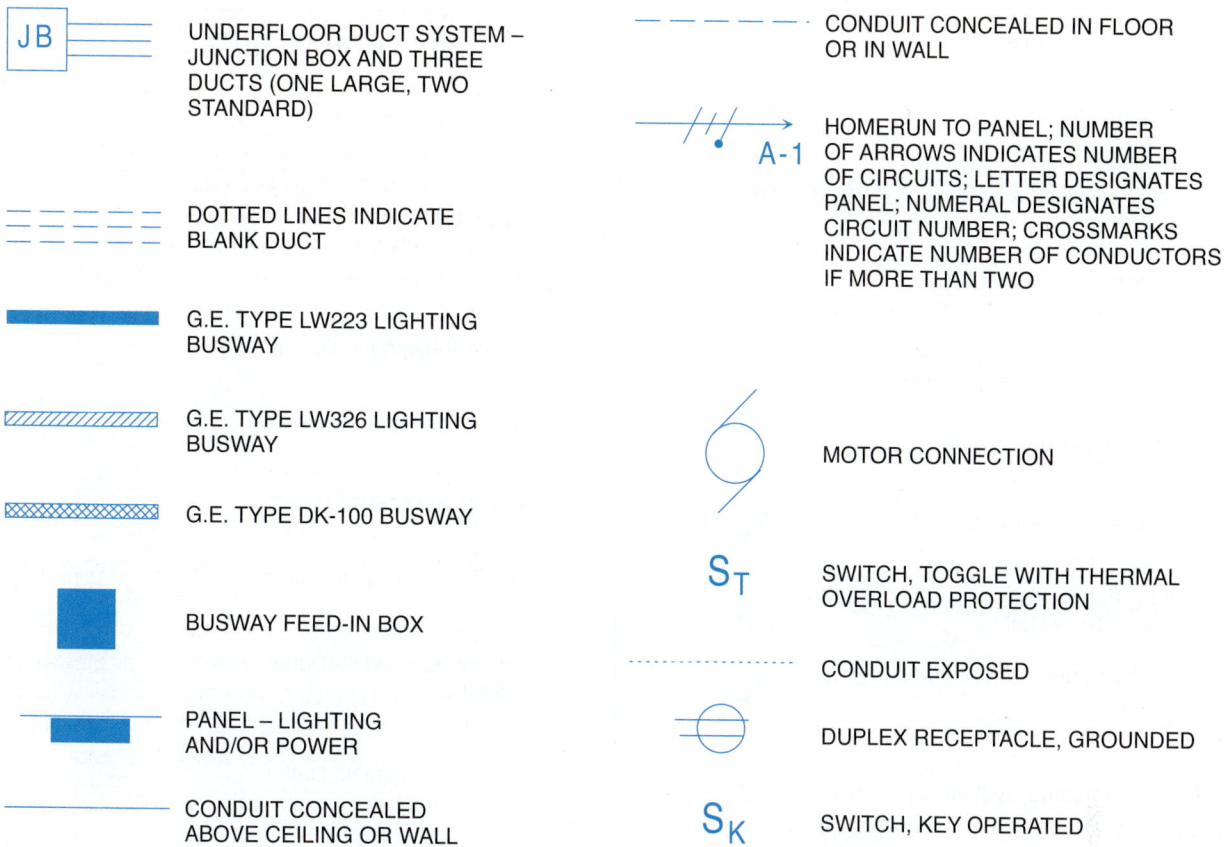

Figure 19 Electrical symbols used by one consulting engineering firm.

110F19.EPS

In dimensioning drawings, the dimensions written on the drawing are the actual dimensions of the building, not the distances that are measured on the drawing. To further illustrate this point, look at the floor plan in *Figure 21*; it is drawn to a scale of ½" = 1'-0". One of the walls is drawn to an actual length of 3½" on the drawing paper, but since the scale is ½" = 1'-0" and since 3½" contains 7 halves of an inch (7 × ½ = 3½"), the dimension shown on the drawing will therefore be 7'-0" on the actual building.

As shown in the previous example, the most common method of reducing all the dimensions (in feet and inches) in the same proportion is to choose a certain distance and let that distance represent one foot. This distance can then be divided into 12 parts, each of which represents an inch. If half inches are required, these twelfths are further subdivided into halves, etc. Now the scale represents the common foot rule with its subdivisions into inches and fractions, except that the scaled foot is smaller than the distance known as a foot and, likewise, its subdivisions are proportionately smaller.

When a measurement is made on the drawing, it is made with the reduced foot rule or scale; when a measurement is made on the building, it is made with the standard foot rule. The most common reduced foot rules or scales used in electrical drawings are the architect's scale and the engineer's scale. Drawings may sometimes be encountered that use a metric scale, but using this scale is similar to using the architect's or engineer's scales.

5.1.0 Architect's Scale

Figure 22 shows two configurations of architect's scales. The one on the top is designed so that 1" = 1'-0", and the one on the bottom has graduations spaced to represent ⅛" = 1'-0".

Note that on the one-inch scale in *Figure 23*, the longer marks to the right of the zero (with a numeral beneath) represent feet. Therefore, the distance between the zero and the numeral 1 equals one foot. The shorter mark between the zero and 1 represents ½ of a foot, or six inches.

SWITCH OUTLETS		RECEPTACLE OUTLETS	
Single Pole Switch	S	Where weatherproof, explosionproof, or other specific types of devices are to be required, use the upper-case subscript letters to specify. For example, weatherproof single or duplex receptacles would have the upper-case WP subscript letters noted alongside the symbol. All outlets must be grounded.	
Double Pole Switch	S_2		
Three-Way Switch	S_3		
Four-Way Switch	S_4	Single Receptacle Outlet	
Key-Operated Switch	S_K	Duplex Receptacle Outlet	
Switch and Fusestat Holder	$S_F H$	Triplex Receptacle Outlet	
Switch and Pilot Lamp	S_P	Quadruplex Receptacle Outlet	
Fan Switch	S_F	Duplex Receptacle Outlet Split Wired	
Switch for Low-Voltage Switching System	S_L	Triplex Receptacle Outlet Split Wired	
Master Switch for Low-Voltage Switching System	S_{LM}	250-Volt Receptacle/Single Phase Use Subscript Letter to Indicate Function (DW - Dishwasher, RA - Range) or Numerals (with explanation in symbols schedule)	
Switch and Single Receptacle	S		
Switch and Duplex Receptacle	S	250-Volt Receptacle/Three Phase	
Door Switch	S_D	Clock Receptacle	Ⓒ
Time Switch	S_T	Fan Receptacle	Ⓕ
Momentary Contact Switch	S_{MC}	Floor Single Receptacle Outlet	
Ceiling Pull Switch	Ⓢ	Floor Duplex Receptacle Outlet	
"Hand-Off-Auto" Control Switch	HOA	Floor Special-Purpose Outlet	*
Multi-Speed Control Switch	M	Floor Telephone Outlet - Public	
Pushbutton	▪	Floor Telephone Outlet - Private	

Use numeral keyed explanation of symbol usage

110F20A.EPS

Figure 20 Recommended electrical symbols (1 of 7).

Example of the use of several floor outlet symbols to identify a 2, 3, or more gang outlet:

Underfloor duct and junction box for triple, double, or single duct system as indicated by the number of parallel lines

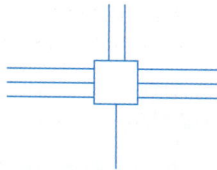

Example of the use of various symbols to identify the location of different types of outlets or connections for underfloor duct or cellular floor systems:

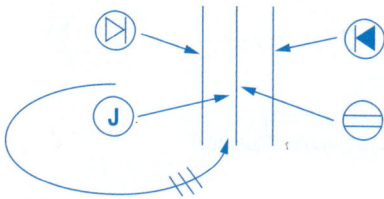

Cellular Floor Heater Duct

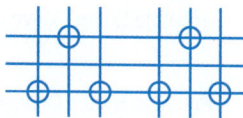

CIRCUITING

Wiring Exposed (not in conduit)	—— E ——
Wiring Concealed in Ceiling or Wall	————————
Wiring Concealed in Floor	— — — — —
Wiring Existing*	··················
Wiring Turned Up	———○
Wiring Turned Down	———●
Branch Circuit Homerun to Panelboard	2 1 ➤➤

Number of arrows indicates number of circuits. (A number at each arrow may be used to identify the circuit number.)**

BUS DUCTS AND WIREWAYS

Trolley Duct***	T T
Busway (Service, Feeder or Plug-in)***	B B
Cable Trough Ladder or Channel***	C C
Wireway***	W W

PANELBOARDS, SWITCHBOARDS AND RELATED EQUIPMENT

Flush Mounted Panelboard and Cabinet***

Surface Mounted Panelboard and Cabinet***

Switchboard, Power Control Center, Unit Substation (Should be drawn to scale)***

Flush Mounted Terminal Cabinet (In small scale drawings the TC may be indicated alongside the symbol)*** TC

Surface Mounted Terminal Cabinet (In small scale drawings the TC may be indicated alongside the symbol)*** TC

Pull Box (Identify in relation to Wiring System Section and Size)

Motor or Other Power Controller May be a starter or contactor***

Externally Operated Disconnection Switch***

Combination Controller and Disconnection Means***

*Note: Use heavy-weight line to identify service and feeders. Indicate empty conduit by notation CO.

**Note: Any circuit without further identification indicates two-wire circuit. For a greater number of wires, indicate with cross lines, e.g.:

3 wires   4 wires, etc.

Neutral and ground wires may be shown longer. Unless indicated otherwise, the wire size of the circuit is the minimum size required by the specification. Identify different functions of wiring system (e.g., signaling system) by notation or other means.

***Identify by Notation or Schedule

Figure 20 Recommended electrical symbols (2 of 7).

110F20B.EPS

POWER EQUIPMENT

Electric Motor (HP as Indicated)	1/4
Power Transformer	
Pothead (Cable Termination)	
Circuit Element e.g., Circuit Breaker	CB
Circuit Breaker	
Fusible Element	
Single-Throw Knife Switch	
Double-Throw Knife Switch	
Ground	
Battery	
Contactor	C
Photoelectric Cell	PE
Voltage Cycles, Phase	EX: 480/60/3
Relay	R
Equipment Connection (as noted)	

REMOTE CONTROL STATIONS FOR MOTORS OR OTHER EQUIPMENT

Pushbutton Station	PB
Float Switch - Mechanical	F
Limit Switch - Mechanical	L
Pneumatic Switch - Mechanical	P
Electric Eye - Beam Source	
Electric Eye - Relay	
Temperature Control Relay Connection (3 Denotes Quantity)	R 3
Solenoid Contrl Valve Connection	S
Pressure Switch Connection	P
Aquastat Connection	A
Vacuum Switch Connection	V
Gas Solenoid Valve Connection	G
Flow Switch Connection	F
Timer Connection	T
Limit Switch Connection	L

LIGHTING OUTLETS

	Ceiling	Wall
Incandescent Fixture (Surface or Pendant)		
Incandescent Fixture with Pull Chain (Surface or Pendant)	PC	PC

110F20C.EPS

Figure 20   Recommended electrical symbols (3 of 7).

		Ceiling	Wall

Exit Light (Surface or Pendant) — ⊗ (Ceiling), ⊗ (Wall)

Blanked Outlet — Ⓑ (Ceiling), Ⓑ (Wall)

Junction Box — Ⓙ (Ceiling), Ⓙ (Wall)

Recessed Incandescent Fixture

Individual Fluorescent Fixture (Surface or Pendant)

Continuous Row Fluorescent Fixture (Surface or Pendant)

Letter indicating controlling switch → A

$\dfrac{1}{100}$ ← Fixture No. ← Wattage

Symbol not needed at each fixture

Bare-Lamp Fluorescent Strip*

ELECTRIC DISTRIBUTION OR LIGHTING SYSTEM, AERIAL

Pole**

Street or Parking Lot Light and Bracket

Transformer**

Primary Circuit**

Secondary Circuit**

Down Guy

Head Guy

Sidewalk Guy

Service Weatherhead**

ELECTRIC DISTRIBUTION OR LIGHTING SYSTEM, UNDERGROUND

Manhole — M

Handhole — H

Transformer Manhole or Vault — TM

Transformer Pad — TP

Underground Direct Burial Cable (Indicate type, size, and number of conductors by notation or schedule.)

Underground Duct Line (Indicate type, size, and number of ducts by cross-section identification of each run by notation or schedule. Indicate type, size, and number of conductors by notation or schedule.)

Street Light Standard Fed From Underground Circuit**

*In the case of continuous-row bare-lamp fluorescent strip above an area-wide diffusing means, show each fixture run using the standard symbol; indicate area of diffusing means and type by light shading and/or drawing notation.

**Identify by Notation or Schedule

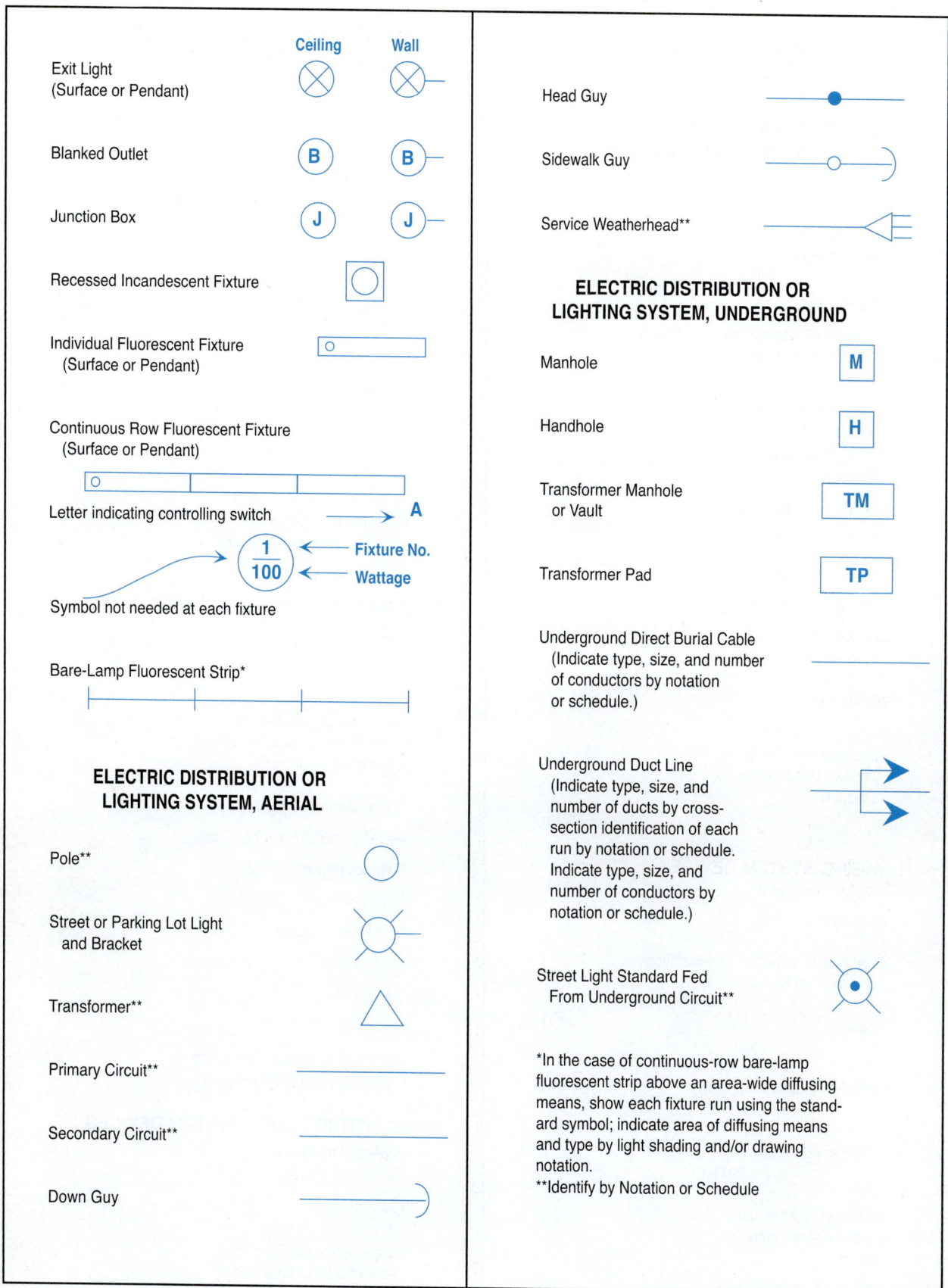

Figure 20 Recommended electrical symbols (4 of 7).

110F20D.EPS

SIGNALING SYSTEM OUTLETS

INSTITUTIONAL, COMMERCIAL, AND INDUSTRIAL OCCUPANCIES

I NURSE CALL SYSTEM DEVICES (Any Type)

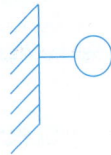

Basic Symbol

(Examples of Individual Item Identification Not a Part of Standard)

Nurses' Annunciator
(Add a number after it as
—①24 to indicate number
of lamps)

Call Station, Single Cord, Pilot Light

Call Station, Double Cord, Microphone Speaker

Corridor Dome Light 1 Lamp

Transformer

Any Other Item On Same System Use Number As Required

II PAGING SYSTEM DEVICES

Basic Symbol

(Examples of Individual Item Identification Not a Part of Standard)

Keyboard

Flush Annunciator

2-Face Annunciator

Any Other Item On Same System Use Numbers As Required

III FIRE ALARM SYSTEM DEVICES (Any Type) Including Smoke and Sprinkler Alarm Devices

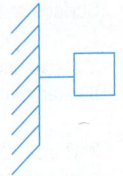

Basic Symbol

(Examples of Individual Item Identification. Not a Part of Standard)

Control Panel

Station

10" Gong

Pre-Signal Chime

Any Other Item On Same System Use Numbers As Required

IV STAFF REGISTER SYSTEM DEVICES (Any Type)

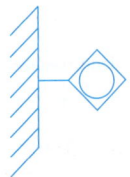

Basic Symbol

(Examples of Individual Item Identification. Not a Part of Standard)

Phone Operators' Register

Entrance Register - Flush

Staff Room Register

Transformer

Any Other Item On Same System Use Numbers As Required

V ELECTRIC CLOCK SYSTEM DEVICES (Any Type)

Basic Symbol

(Examples of Individual Item Identification. Not a Part of Standard)

Figure 20   Recommended electrical symbols (5 of 7).

110F20E.EPS

Master Clock ⬡ 1

12" Secondary - Flush ⬡ 2

12" Double Dial - Wall Mounted ⬡ 3

18" Skeleton Dial ⬡ 4

Any Other Item On Same System
Use Numbers As Required ⬡ 5

VI   PUBLIC TELEPHONE SYSTEM DEVICES

Basic Symbol

(Examples of Individual Item
Identification. Not a Part of
Standard)

Switchboard ◀ 1

Desk Phone ◀ 2

Any Other Item On Same
System Use Numbers As
Required ◀ 3

VII   PRIVATE TELEPHONE SYSTEM DEVICES
(Any Type)

Basic Symbol

(Examples of Individual Item
Identification. Not a Part of
Standard)

Switchboard ◁ 1

Wall Phone ◁ 2

Any Other Item On Same System
Use Numbers As Required ◁ 3

VIII   WATCHMAN SYSTEM DEVICES
(Any Type)

Basic Symbol

(Examples of Individual Item
Identification. Not a Part of
Standard)

Central Station ⬠ 1

Key Station ⬠ 2

Any Other Item On Same System
Use Numbers As Required ⬠ 3

IX   SOUND SYSTEM

Basic Symbol

(Examples of Individual Item
Identification. Not a Part of
Standard)

Amplifier ◁ 1

Microphone ◁ 2

Interior Speaker ◁ 3

Exterior Speaker ◁ 4

Any Other Item On Same System
Use Numbers As Required ◁ 5

X   OTHER SIGNAL SYSTEM DEVICES

Basic Symbol

(Examples of Individual Item
Identification. Not a Part of
Standard)

Buzzer ▢ 1

Bell ▢ 2

Pushbutton ▢ 3

Annunciator ⬤ 4

Any Other Item On Same System
Use Numbers As Required ⬤ 5

Figure 20   Recommended electrical symbols (6 of 7).

110F20F.EPS

RESIDENTIAL OCCUPANCIES

Signaling system symbols for use in identifying standardized residential-type signal system items on residential drawings where a descriptive symbol list is not included on the drawing. When other signal system items are to be identified, use the above basic symbols for such items together with a descriptive symbol list.

Pushbutton	
Buzzer	
Bell	
Combination Bell - Buzzer	
Chime	CH
Annunciator	
Electric Door Opener	D
Maid's Signal Plug	M
Interconnection Box	
Bell-Ringing Transformer	BT
Outside Telephone	
Interconnecting Telephone	
Television Outlet	TV

110F20G.EPS

Figure 20   Recommended electrical symbols (7 of 7).

Referring again to *Figure 23*, look at the marks to the left of the zero. The numbered marks are spaced three scaled inches apart and have the numerals 0, 3, 6, and 9 for use as reference points. The other lines of the same length also represent scaled inches, but are not marked with numerals. In use, you can count the number of long marks to the left of the zero to find the number of inches, but after some practice, you will be able to tell the exact measurement at a glance. For example, the measurement A represents five inches because it is the fifth inch mark to the left of the zero; it is also one inch mark short of the six-inch line on the scale.

The lines that are shorter than the inch line are the half-inch lines. On smaller scales, the basic unit is not divided into as many divisions. For example, the smallest subdivision on some scales represents two inches.

5.1.1 Types of Architect's Scales

Architect's scales are available in several types, but the most common include the triangular scale (*Figure 24*) and the flat scale. The quality of architect's scales also varies from cheap plastic scales (costing a dollar or two) to high-quality wooden-laminated tools that are calibrated to precise standards.

The triangular scale is frequently found in drafting and estimating departments or engineering and electrical contracting firms, while the flat scales are more convenient to carry on the job site.

Triangular architect's scales have 12 different scales—two on each edge—as follows:

- Common foot rule (12 inches)
- $\frac{1}{16}$" = 1'–0"
- $\frac{3}{32}$" = 1'–0"
- $\frac{3}{16}$" = 1'–0"
- $\frac{1}{8}$" = 1'–0"
- $\frac{1}{4}$" = 1'–0"
- $\frac{3}{8}$" = 1'–0"
- $\frac{3}{4}$" = 1'–0"
- 1" = 1'–0"
- $\frac{1}{2}$" = 1'–0"
- $1\frac{1}{2}$" = 1'–0"
- 3" = 1'–0"

Two separate scales on one face may seem confusing at first, but after some experience, reading these scales becomes second nature.

In all but one of the scales on the triangular architect's scale, each face has one of the scales placed opposite to the other. For example, on the one-inch face, the one-inch scale is read from left to right, starting from the zero mark. The half-inch scale is read from right to left, again starting from the zero mark.

PUMP HOUSE FLOOR PLAN
½" = 1'–0"

The distance between the arrowheads to the left measures 3½" on the drawing, but since the drawing is made to a scale of ½" = 1'–0", this measurement actually represents 7'–0".

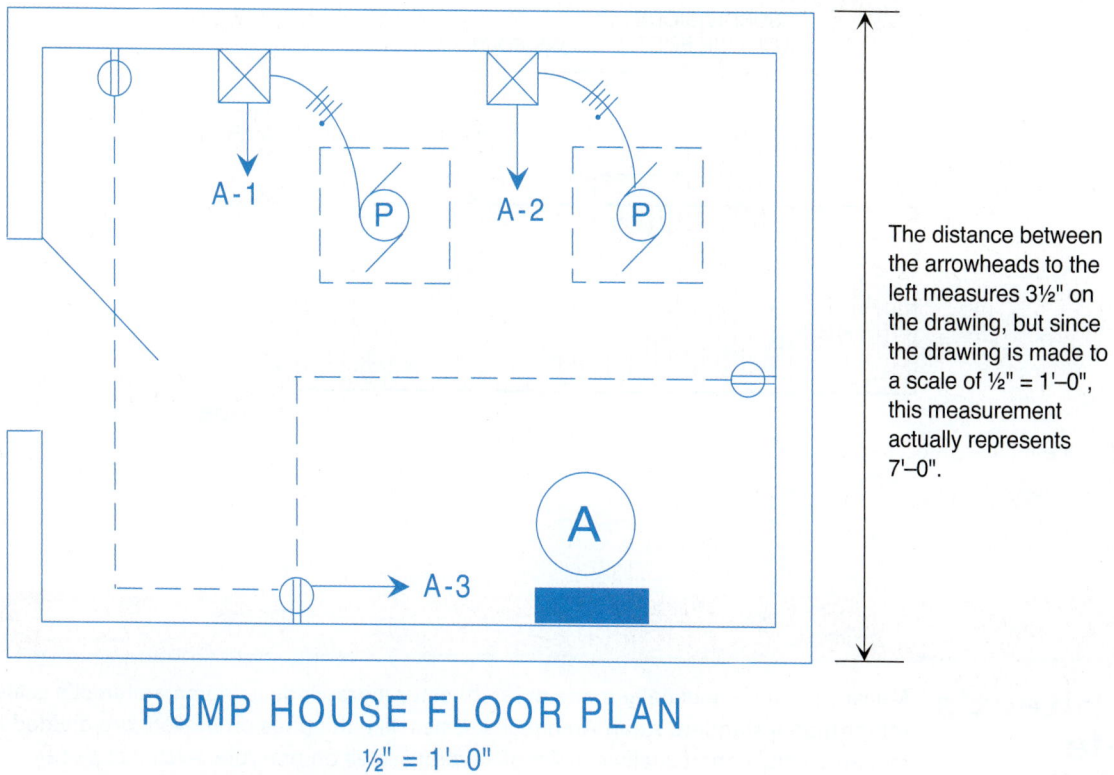

Figure 21   Typical floor plan showing drawing scale.

110F21.EPS

On the remaining foot-rule scale (⅟₁₆" = 1'–0") each ⅟₁₆" mark on the scale represents one foot.

Figure 25 shows all the scales found on the triangular architect's scale.

The flat architect's scale shown in *Figure 26* is ideal for workers on most projects. It is easily and conveniently carried in the shirt pocket, and the four scales (⅛", ¼", ½", and 1") are adequate for the majority of projects that will be encountered.

The partial floor plan shown in *Figure 26* is drawn to a scale of ⅛" = 1'–0". The dimension in question is found by placing the ⅛" architect's scale on the drawing and reading the figures. It can be seen that the dimension reads 24'–6".

Every drawing should have the scale to which it is drawn plainly marked on it as part of the drawing title. However, it is not uncommon to have several different drawings on one blueprint sheet—all with different scales. Therefore, always check the scale of each different view found on a drawing sheet.

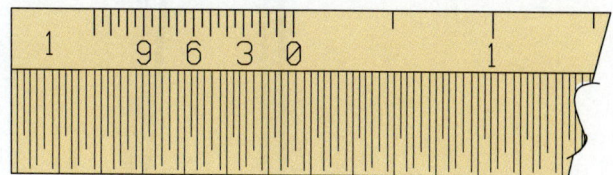

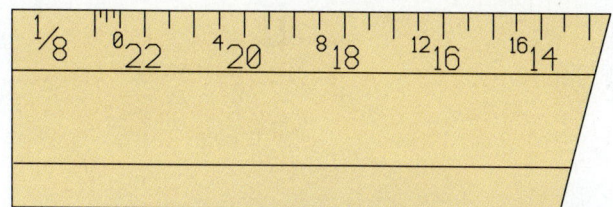

1" = 1'–0"

⅛" = 1'–0"

110F22.EPS

Figure 22   Two different configurations of architect's scales.

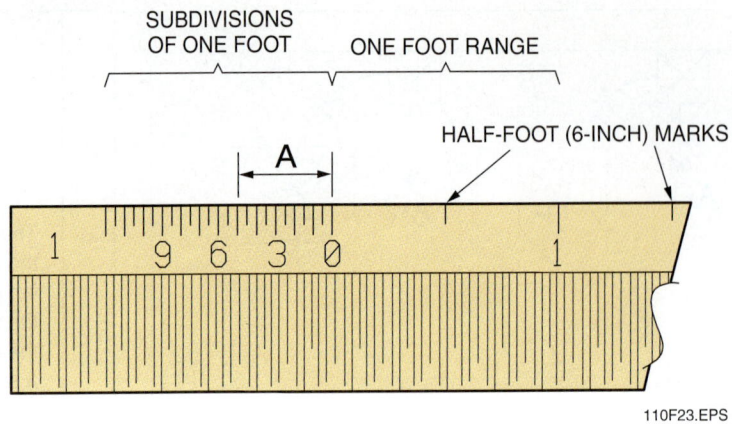

Figure 23 One-inch architect's scale.

On Site

Architect's Scale

Measurements are usually made on architectural drawings using an architect's scale rather than a standard ruler. Architect's scales, like the ones on the left, are divided into feet and inches and usually consist of several scales on one rule. Architect's scales also come in other forms such as tapes or with wheels, like the one shown on the right.

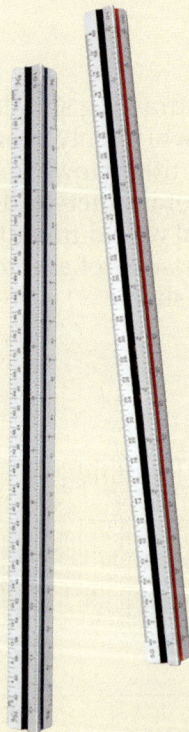

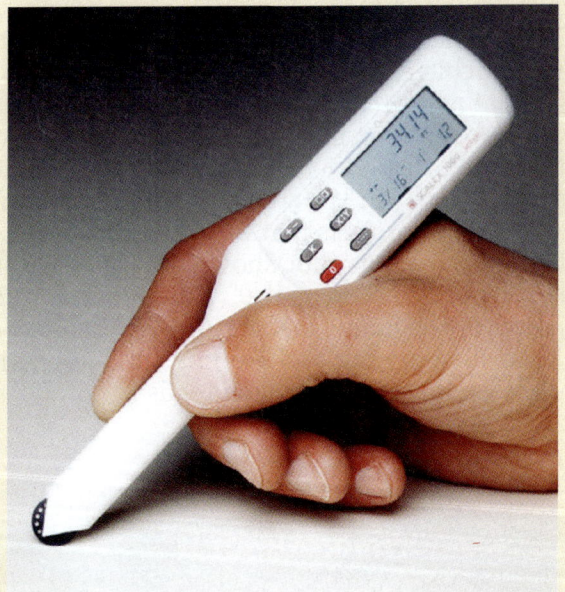

NCCER — *Contren® Learning Series* 26110-11

Figure 24 Typical triangular architect's scale.

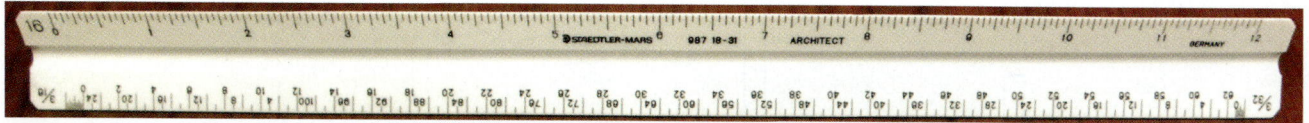

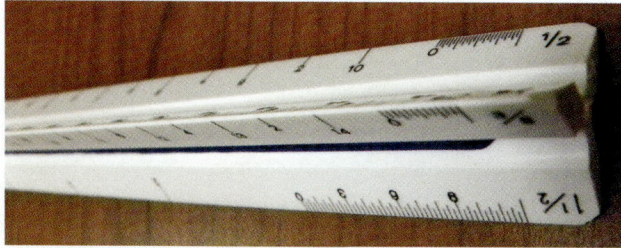

Figure 25 Various scales on a triangular architect's scale.

5.2.0 Engineer's Scale

The civil engineer's scale is used in basically the same manner as the architect's scale, with the principal difference being that the graduations on the engineer's scale are decimal units rather than feet as on the architect's scale.

The engineer's scale is used by placing it on the drawing with the working edge away from the user. The scale is then aligned in the direction of the required measurement. Then, by looking down at the scale, the dimension is read.

Civil engineer's scales commonly show the following graduations:

- 1" = 10 units
- 1" = 20 units
- 1" = 30 units
- 1" = 40 units
- 1" = 60 units
- 1" = 80 units
- 1" = 100 units

The purpose of this scale is to transfer the relative dimensions of an object to the drawing or vice versa. It is used mainly on site plans to determine distances between property lines, manholes, duct runs, direct-burial cable runs, and the like.

Site plans are drawn to scale using the engineer's scale rather than the architect's scale. On small lots, a scale of 1 inch = 10 feet or 1 inch = 20 feet is used. For a 1:10 scale, this means that one inch (the actual measurement on the drawing) is equal to 10 feet on the land itself.

On larger drawings, where a large area must be covered, the scale could be 1 inch = 100 feet or 1 inch = 1,000 feet, or any other integral power of 10. On drawings with the scale in multiples of 10, the engineer's scale marked 10 is used. If the scale is 1 inch = 200 feet, the engineer's scale marked 20 is used, and so on.

Although site plans appear reduced in scale, depending on the size of the object and the size of the drawing sheet to be used, the actual dimensions must be shown on the drawings at all times. When you are reading the drawing plans to scale, think of each dimension in its full size and not in the reduced scale it happens to be on the drawing (*Figure 27*).

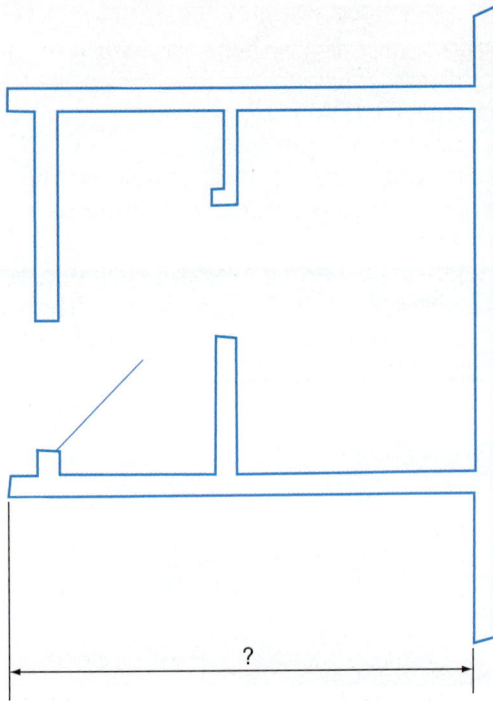

SCALE: ⅛" = 1'–0"

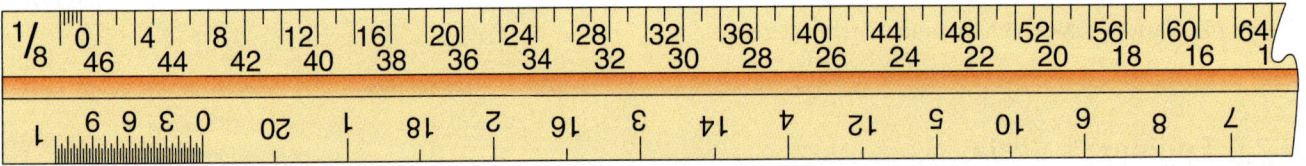

110F26.EPS

Figure 26   Using the ⅛" architect's scale to determine the dimensions on a drawing.

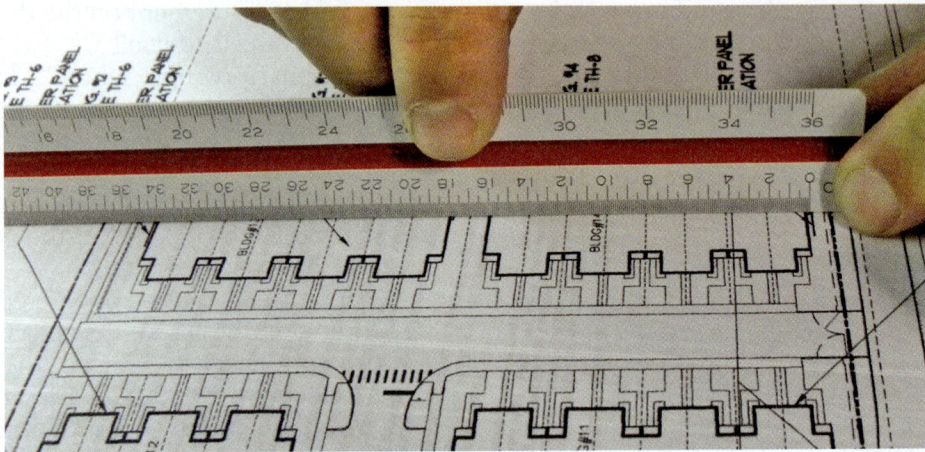

110F27.EPS

Figure 27   Practical use of the engineer's scale.

5.3.0 Metric Scale

Metric scales are calibrated in units of 10 (*Figure 28*). The two common length measurements used in the metric scale on architectural drawings are the meter and the millimeter, the millimeter being ¹⁄₁,₀₀₀ of a meter. On drawings drawn to scales between 1:1 and 1:100, the millimeter is typically used. On drawings drawn to scales between 1:200 and 1:2,000, the meter is generally used. Many contracting firms that deal in international trade have adopted a dual-dimensioning system expressed in both metric and English symbols. Drawings prepared for government projects may also require metric dimensions. A metric conversion chart is provided in the Appendix.

6.0.0 ANALYZING ELECTRICAL DRAWINGS

The most practical way to learn how to read electrical construction documents is to analyze an existing set of drawings prepared by consulting or industrial engineers.

Engineers or electrical designers are responsible for the complete layout of electrical systems for most projects. Electrical drafters then transform the engineer's designs into working drawings, using either manual drafting instruments or computer-aided design (CAD) systems. The following is a brief outline of what usually takes place in the preparation of electrical design and working drawings:

- The engineer meets with the architect and owner to discuss the electrical needs of the building or project and to discuss various recommendations made by all parties.
- After that, an outline of the architect's floor plan is laid out.
- The engineer then calculates the required power and lighting outlets for the project; these are later transferred to the working drawings.
- All communications and alarm systems are located on the floor plan, along with lighting and power panelboards.
- Circuit calculations are made to determine wire size and overcurrent protection.

- The main electric service and related components are determined and shown on the drawings.
- Schedules are then placed on the drawings to identify various pieces of equipment.
- Wiring diagrams are made to show the workers how various electrical components are to be connected.
- A legend or electrical symbol list is drafted and shown on the drawings to identify all symbols used to indicate electrical outlets or equipment.
- Various large-scale electrical details are included, if necessary, to show exactly what is required of the electricians.
- Written specifications are then made to give a description of the materials and installation methods.

6.1.0 Development of Site Plans

In general practice, it is usually the owner's responsibility to furnish the architect/engineer with property and topographic surveys, which are made by a certified land surveyor or civil engineer. These surveys show:

- All property lines
- Existing public utilities and their location on or near the property (e.g., electrical lines, sanitary sewer lines, gas lines, water-supply lines, storm sewers, manholes, telephone lines, etc.)

A land surveyor does the property survey from information obtained from a deed description of the property. A property survey shows only the property lines and their lengths, as if the property were perfectly flat.

The topographic survey shows both the property lines and the physical characteristics of the land by using contour lines, notes, and symbols. The physical characteristics may include:

- The direction of the land slope
- Whether the land is flat, hilly, wooded, swampy, high, or low, and other features of its physical nature

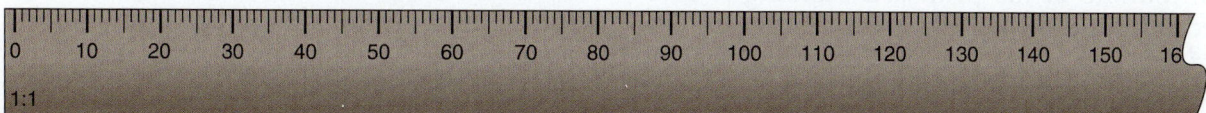

110F28.EPS

Figure 28   Typical metric scale.

All of this information is necessary so that the architect can properly design a building to fit the property. The electrical engineer also needs this information to locate existing electrical utilities and to route the new service to the building, provide outdoor lighting and circuits, etc.

Electrical site work is sometimes shown on the architect's plot plan. However, when site work involves many trades and several utilities (e.g., gas, telephone, electric, television, water, and sewage), it can become confusing if all details are shown on one drawing sheet. In cases like these, it is best to have a separate drawing devoted entirely to the electrical work, as shown in *Figure 29*. This project is an office/warehouse building for Virginia Electric, Inc. The electrical drawings consist of four 24" × 36" drawing sheets, along with a set of written specifications, which will be discussed later in this module.

The electrical site or plot plan shown in *Figure 29* has the conventional architect's and engineer's title blocks in the lower right-hand corner of the drawing. These blocks identify the project and project owners, the architect, and the engineer. They also show how this drawing sheet relates to the entire set of drawings. Note the engineer's professional stamp of approval to the left of the engineer's title block. Similar blocks appear on all four of the electrical drawing sheets.

When examining a set of electrical drawings for the first time, always look at the area around the title block. This is where most revision blocks or revision notes are placed. If revisions have been made to the drawings, make certain that you have a clear understanding of what has taken place before proceeding with the work.

Refer again to the drawing in *Figure 29* and note the north arrow in the upper left corner. A north arrow shows the direction of true north to help you orient the drawing to the site. Look directly down from the north arrow to the bottom of the page and notice the drawing title, *Plot Utilities*. Directly beneath the drawing title you can see that the drawing scale of 1" = 30' is shown. This means that each inch on the drawing represents 30 feet on the actual job site. This scale holds true for all drawings on the page unless otherwise noted.

An outline of the proposed building is indicated on the drawing along with a callout, *Proposed Bldg. Fin. Flr. Elev. 590.0*. This means that the finished floor level of the building is to be 590 feet above sea level, which in this part of the country will be about two feet above finished grade around the building. This information helps the electrician locate conduit sleeves and stub-ups to the correct height before the finished concrete floor is poured.

The shaded area represents asphalt paving for the access road, drives, and parking lot. Note that the access road leads into a highway, which is designated Route 35. This information further helps workers to orient the drawing to the building site.

Existing manholes are indicated by a solid circle, while an open circle is used to show the position of the five new pole-mounted lighting fixtures that are to be installed around the new building. Existing power lines are shown with a light solid line with the letter E placed at intervals along the line. The new underground electric service is shown in the same way, except the lines are somewhat wider and darker on the drawing. Note that this new high-voltage cable terminates into a padmount transformer near the proposed building. New telephone lines are similar except the letter T is used to identify the telephone lines.

The direct-burial underground cable supplying the exterior lighting fixtures is indicated with dashed lines on the drawing—shown connecting the open circles. A homerun for this circuit is also shown to a time clock.

The manhole detail shown to the right of the north arrow may seem to serve very little purpose on this drawing since the manholes have already been installed. However, the dimensions and details of their construction will help the electrical contractor or supervisor to better plan the pulling of the high-voltage cable. The same is true of the cross section shown of the duct bank. The electrical contractor knows that three empty ducts are available if it is discovered that one of them is damaged when the work begins.

Although the electrical work will not involve working with gas, the main gas line is shown on the electrical drawing to let the electrical workers know its approximate location while they are installing the direct-burial conductors for the exterior lighting fixtures.

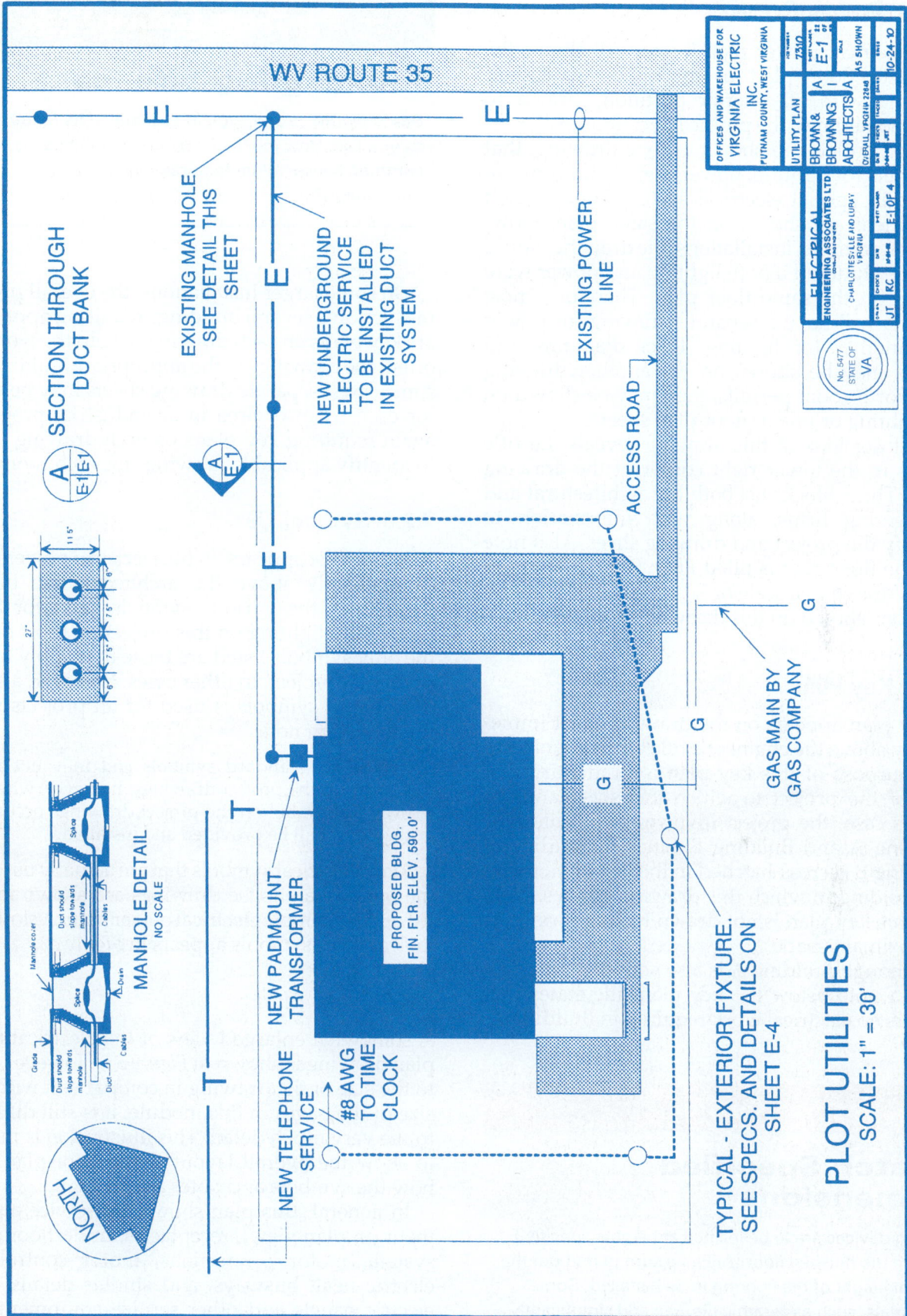

Figure 29 Typical electrical site plan.

The drawing contains the following labels:

WV ROUTE 35

SECTION THROUGH DUCT BANK

A
E-1 | E-1

EXISTING MANHOLE SEE DETAIL THIS SHEET

NEW UNDERGROUND ELECTRIC SERVICE TO BE INSTALLED IN EXISTING DUCT SYSTEM

A
E-1

EXISTING POWER LINE

ACCESS ROAD

G

G

G

GAS MAIN BY GAS COMPANY

TYPICAL - EXTERIOR FIXTURE. SEE SPECS. AND DETAILS ON SHEET E-4

PLOT UTILITIES
SCALE: 1" = 30'

PROPOSED BLDG.
FIN. FLR. ELEV. 590.0'

NEW PADMOUNT TRANSFORMER

NEW TELEPHONE SERVICE

#8 AWG TO TIME CLOCK

MANHOLE DETAIL
NO SCALE

Manhole cover
Spice
Duct should slope towards manhole
Grade
Duct should slope towards manhole
Spice
Duct
Cables
Cables
Drain

NORTH

115'
27"
6" 7.5" 7.5" 6"

110F29.EPS

OFFICES AND WAREHOUSE FOR VIRGINIA ELECTRIC INC.
PUTNAM COUNTY, WEST VIRGINIA

UTILITY PLAN

BROWN & BROWNING ARCHITECTS A

CHARLOTTESVILLE AND LURAY VIRGINIA

ELECTRICAL ENGINEERING ASSOCIATES LTD

No. 5477 STATE OF VA

JT | RC

E-1 OF 4

7310

E-1

AS SHOWN

10-24-10

Basic Electrical Construction Drawings

31

7.0.0 POWER PLANS

The electrical power plan (*Figure 30*) shows the complete floor plan of the office/warehouse building with all interior partitions drawn to scale. Sometimes, the physical locations of all wiring and outlets are shown on one drawing; that is, outlets for lighting, power, signal and communications, special electrical systems, and related equipment are shown on the same plan. However, on complex installations, the drawing would become cluttered if both lighting and power were shown on the same floor plan. Therefore, most projects will have a separate drawing for power and another for lighting. Riser diagrams and details may be shown on yet another drawing sheet, or if room permits, they may be shown on the lighting or power floor plan sheets.

A closer look at this drawing reveals the title blocks in the lower right corner of the drawing sheet. These blocks list both the architectural and engineering firms, along with information to identify the project and drawing sheet. Also note that the floor plan is titled *Floor Plan "B"—Power* and is drawn to a scale of $\frac{1}{8}" = 1'-0"$. There are no revisions shown on this drawing sheet.

7.1.0 Key Plan

A key plan appears on the drawing sheet immediately above the engineer's title block (*Figure 31*). The purpose of this key plan is to identify that part of the project to which this sheet applies. In this case, the project involves two buildings: Building A and Building B. Since the outline of Building B is cross-hatched in the key plan, this is the building to which this drawing applies. Note that this key plan is not drawn to scale—only its approximate shape.

Although Building A is also shown on this key plan, a note below the key plan title states that there is no electrical work required in Building A.

On some larger installations, the overall project may involve several buildings requiring appropriate key plans on each drawing to help the workers orient the drawings to the appropriate building. In some cases, separate drawing sheets may be used for each room or area in an industrial project—again requiring key plans on each drawing sheet to identify applicable drawings for each room.

7.2.0 Symbol List

A symbol list appears on the electrical power plan (immediately above the architect's title block) to identify the various symbols used for both power and lighting on this project. In most cases, the only symbols listed are those that apply to the particular project. In other cases, however, a standard list of symbols is used for all projects with the following note:

> "These are standard symbols and may not all appear on the project drawings; however, wherever the symbol on the project drawings occurs, the item shall be provided and installed."

Only electrical symbols that are actually used for the office/warehouse drawings are shown in the list on the example electrical power plan. A close-up look at these symbols appears in *Figure 32*.

7.3.0 Floor Plan

A somewhat enlarged view of the electrical floor plan drawing is shown in *Figure 33*. However, due to the size of the drawing in comparison with the size of the pages in this module, it is still difficult to see very much detail. This illustration is meant to show the overall layout of the floor plan and how the symbols and notes are arranged.

In general, this plan shows the service equipment (in plan view), receptacles, underfloor duct system, motor connections, motor controllers, electric heat, busways, and similar details. The electric panels and other service equipment are drawn close to scale. The locations of other electrical outlets and similar components are only

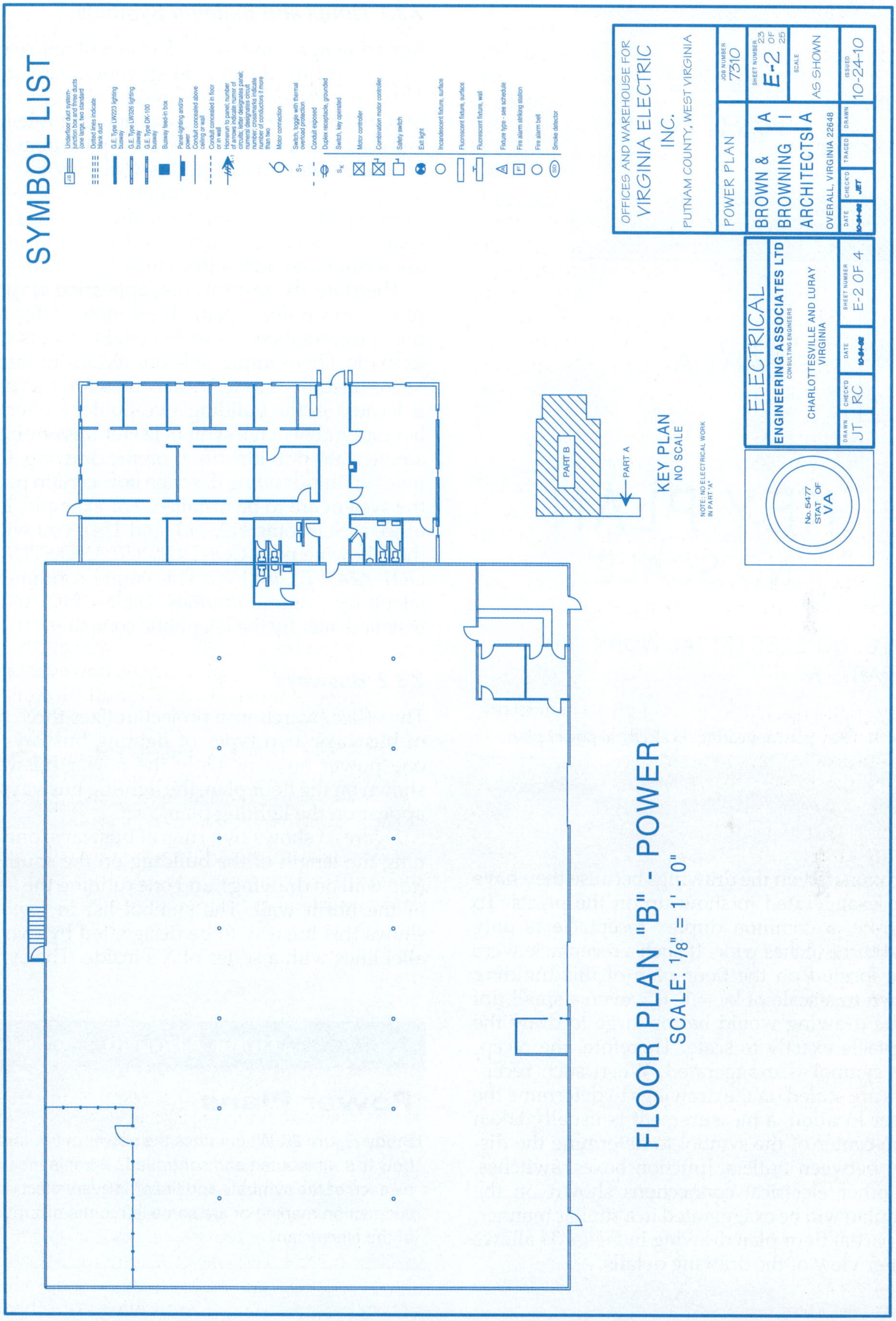

SYMBOL LIST

Underfloor duct system—
junction box and three ducts
(one large, two standard)

Dotted lines indicate
blank duct

G.E. Type LW223 lighting
busway

G.E. Type LW226 lighting
busway

G.E. Type DK-100
busway

Busway feed-in box

Panel—lighting and/or
power

Conduit concealed above
ceiling or wall

Conduit concealed in floor
or in wall

Homerun to panel; number
of arrows indicate number of
circuits; letter designates panel;
numeral designates circuit
number; crossmarks indicate
number of conductors if more
than two

Motor connection

Switch, toggle with thermal
overload protection

Conduit exposed

Duplex receptacle, grounded

Switch, key operated

Motor controller

Combination motor controller

Safety switch

Exit light

Incandescent fixture, surface

Fluorescent fixture, surface

Fluorescent fixture, wall

Fixture type - see schedule

Fire alarm striking station

Fire alarm bell

Smoke detector

FLOOR PLAN "B" - POWER
SCALE: 1/8" = 1' - 0"

KEY PLAN
NO SCALE

PART B

PART A

NOTE: NO ELECTRICAL WORK
IN PART "A"

OFFICES AND WAREHOUSE FOR
VIRGINIA ELECTRIC INC.
PUTNAM COUNTY, WEST VIRGINIA

POWER PLAN

BROWN & BROWNING ARCHITECTS

OVERALL, VIRGINIA 22648

DATE 10-24-09   CHECK'D JET   TRACED

JOB NUMBER 7310
SHEET NUMBER E-2   25 OF 25
SCALE AS SHOWN
ISSUED 10-24-10

ELECTRICAL
ENGINEERING ASSOCIATES LTD
CONSULTING ENGINEERS
CHARLOTTESVILLE AND LURAY
VIRGINIA

DRAWN JT   CHECK'D RC   SHEET NUMBER E-2 OF 4   DATE 10-24-09

No. 5477
STAT OF
VA

110F30.EPS

Figure 30   Electrical power plan.

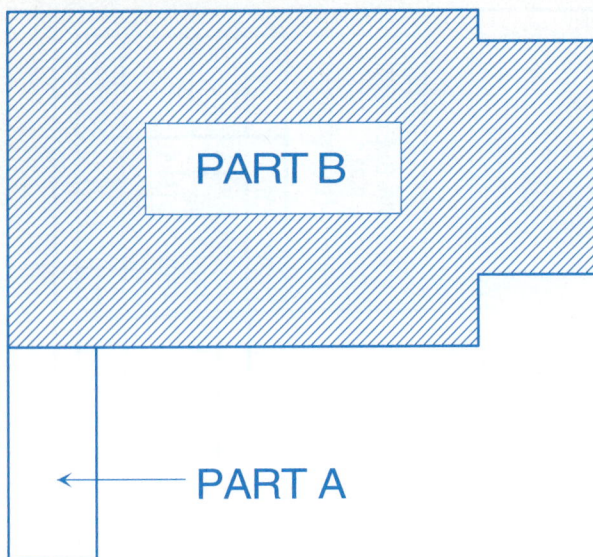

PART B

PART A

KEY PLAN
NO SCALE

NOTE: NO ELECTRICAL WORK
IN PART "A"

110F31.EPS

Figure 31   Key plan appearing on electrical power plan.

approximated on the drawings because they have to be exaggerated to show up on the prints. To illustrate, a common duplex receptacle is only about three inches wide. If such a receptacle were to be located on the floor plan of this building (drawn to a scale of $\frac{1}{8}$" = 1'–0"), even a small dot on the drawing would be too large to draw the receptacle exactly to scale. Therefore, the receptacle symbol is exaggerated. When such receptacles are scaled on the drawings to determine the proper location, a measurement is usually taken to the center of the symbol to determine the distance between outlets. Junction boxes, switches, and other electrical connections shown on the floor plan will be exaggerated in a similar manner. The partial floor plan drawing in *Figure 34* allows a better view of the drawing details.

7.3.1 Notes and Building Symbols

Referring again to *Figure 33*, you will notice numbers placed inside an oval symbol in each room. These numbered ovals represent the room name or type and correspond to a room schedule in the architectural drawings. For example, room number 112 is designated as the lobby in the room schedule (not shown), room number 113 is designated as office No. 1, etc. On some drawings, these room symbols are omitted and the room names are written out on the drawings.

There are also several notes appearing at various places on the floor plan. These notes offer additional information to clarify certain aspects of the drawing. For example, only one electric heater is to be installed by the electrical contractor; this heater is located in the building's vestibule. Rather than have a symbol in the symbol list for this one heater, a note is used to identify it on the drawing. Other notes on this drawing describe how certain parts of the system are to be installed. For example, in the office area (rooms 112, 113, and 114), you will see the following note: *CONDUIT UP AND STUBBED OUT ABOVE CEILING.* This empty conduit is for telephone/communications cables that will be installed later by the telephone company.

7.3.2 Busways

The office/warehouse project utilizes three types of busways: two types of lighting busways and one power busway. Only the power busway is shown on the floor plan; the lighting busways will appear on the lighting plan.

Figure 33 shows two runs of busways: one running the length of the building on the south end (top wall on drawing), and one running the length of the north wall. The symbol list in *Figure 32* shows this busway to be designated by two parallel lines with a series of X's inside. The symbol

<table>
<tr><td colspan="2">**Think About It**</td></tr>
<tr><td colspan="2">**Power Plans**

Study *Figure 33*. Where does the power enter, and how is it distributed and controlled? What is meant by each of the symbols and lines? Is every electrical connection marked or are some left to the discretion of the electrician?</td></tr>
</table>

Figure 32   Sample electrical symbols list.

JB	Underfloor duct system – junction box and three ducts (one large, two standard)
≡≡≡	Dotted lines indicate blank duct
▬▬▬	G.E. Type LW223 lighting busway
▭▭	G.E. Type LW326 lighting busway
▨▨▨	G.E. Type DK-100 busway
■	Busway feed-in box
▬	Panel-lighting and/or power
———	Conduit concealed above ceiling or wall
– – –	Conduit concealed in floor or in wall
⤳ A-1	Homerun to panel; number of arrows indicate number of circuits; letter designates panel; numeral designates circuit number; crossmarks indicate number of conductors if more than two
⟡	Motor connection
S_T	Switch, toggle with thermal overload protection

·········	Conduit exposed
⊖	Duplex receptacle, grounded
S_K	Switch, key operated
⊠	Motor controller
⊠⊣	Combination motor controller
☐⊣	Safety switch
⊗	Exit light
○	Incandescent fixture, surface
▭	Fluorescent fixture, surface
▭	Fluorescent fixture, wall
Ⓐ	Fixture type – see schedule
F	Fire alarm striking station
○	Fire alarm bell
SD	Smoke detector

110F32.EPS

list further describes the busway as General Electric Type DK-100. These busways are fed from the main distribution panel (circuits MDP-1 and MDP-2) through GE No. DHIBBC41 tap boxes.

The *NEC*® defines a busway as a grounded metal enclosure containing factory-mounted, bare or insulated conductors, which are usually copper or aluminum bars, rods, or tubes.

The relationship of the busway and hangers to the building construction should be checked prior to commencing the installation so that any problems due to space conflicts, inadequate or inappropriate supporting structure, openings through walls, etc., are worked out in advance so as not to incur lost time.

For example, the drawings and specifications may call for the busway to be suspended from brackets clamped or welded to steel columns. However, the spacing of the columns may be such that additional supplementary hanger rods suspended from the ceiling or roof structure may be necessary for the adequate support of the busway. To offer more assistance to workers on the office/warehouse project, the engineer may also provide an additional drawing that shows how the busway is to be mounted.

Other details that appear on the floor plan in *Figure 34* include the general arrangement of the underfloor duct system, junction boxes and feeder conduit for the underfloor duct system, and plan views of the service and telephone equipment, along with duplex receptacle outlets. A note on the drawing requires all receptacles in the toilets to be provided with ground fault circuit interrupter (GFCI) protection. The letters EWC next to the receptacle in the vestibule designate this receptacle for use with an electric water cooler.

On Site

Understanding Contact Symbols

When a drawing shows normally open or normally closed contacts, the word *normally* refers to the condition of the contacts in their de-energized or shelf state.

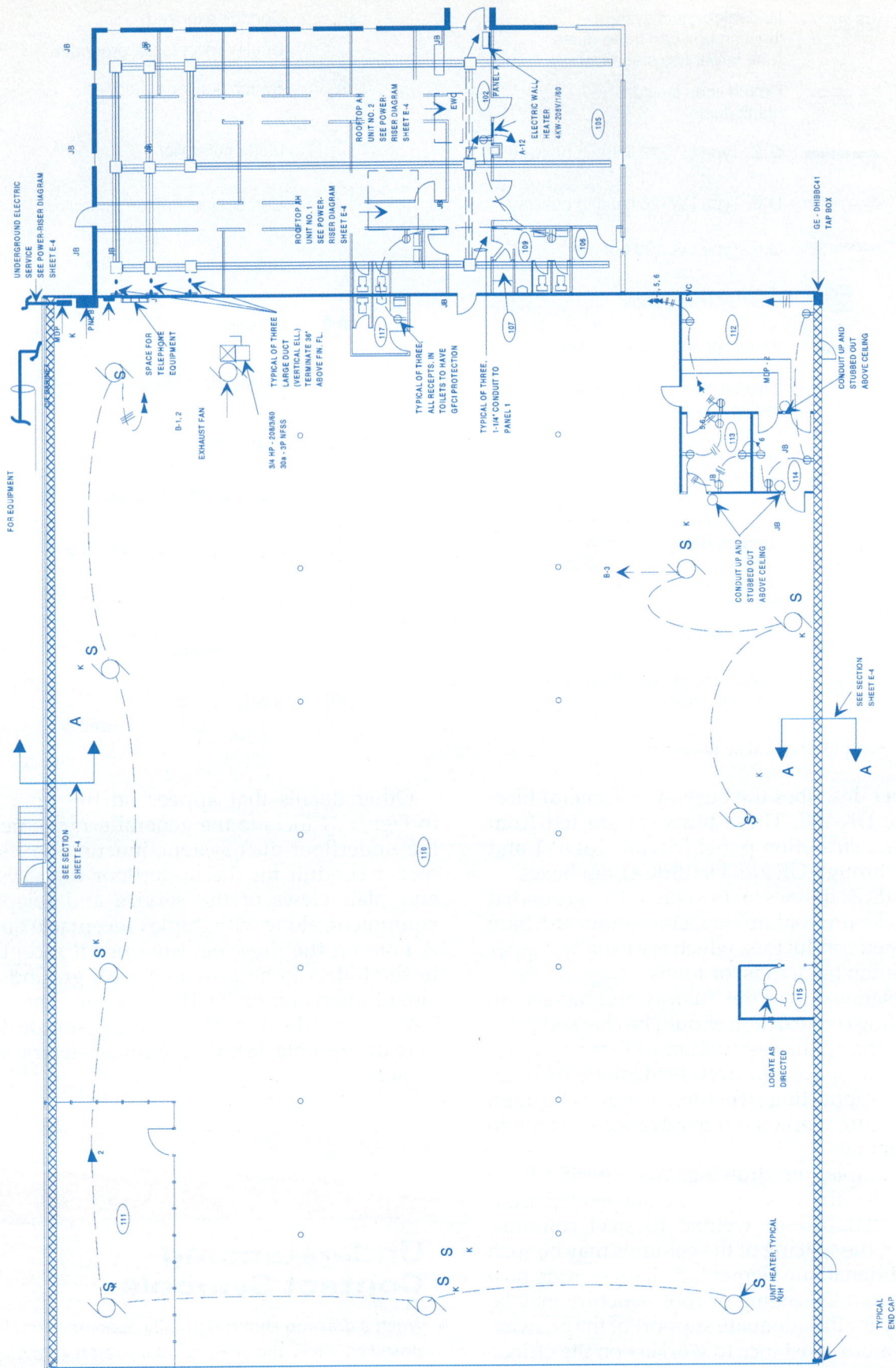

Figure 33   Floor plan for an office/warehouse building.

110F33.EPS

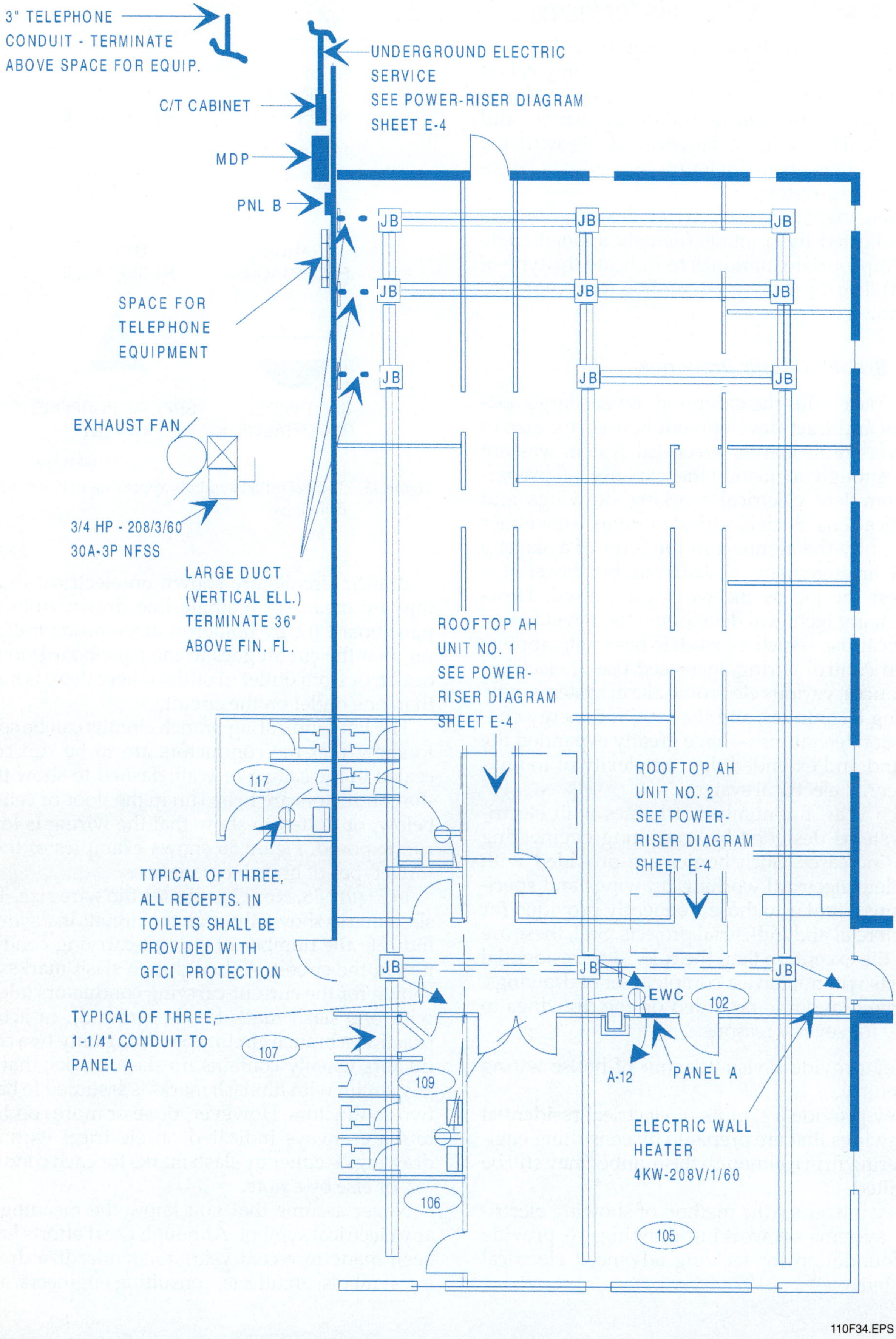

3" TELEPHONE
CONDUIT - TERMINATE
ABOVE SPACE FOR EQUIP.

UNDERGROUND ELECTRIC
SERVICE
SEE POWER-RISER DIAGRAM
SHEET E-4

C/T CABINET

MDP

PNL B

JB   JB   JB

JB   JB   JB

JB   JB   JB

SPACE FOR
TELEPHONE
EQUIPMENT

EXHAUST FAN

3/4 HP - 208/3/60
30A-3P NFSS

LARGE DUCT
(VERTICAL ELL.)
TERMINATE 36"
ABOVE FIN. FL.

ROOFTOP AH
UNIT NO. 1
SEE POWER-
RISER DIAGRAM
SHEET E-4

ROOFTOP AH
UNIT NO. 2
SEE POWER-
RISER DIAGRAM
SHEET E-4

117

TYPICAL OF THREE,
ALL RECEPTS. IN
TOILETS SHALL BE
PROVIDED WITH
GFCI PROTECTION

JB   JB   JB

EWC   102

TYPICAL OF THREE,
1-1/4" CONDUIT TO
PANEL A

107

109

A-12   PANEL A

ELECTRIC WALL
HEATER
4KW-208V/1/60

106

105

110F34.EPS

Figure 34   Partial floor plan for office/warehouse building.

7.4.0 Branch Circuit Layout for Power

The point at which electrical equipment is connected to the wiring system is commonly called an outlet. There are many classifications of outlets: lighting, receptacle, motor, appliance, and so forth. This section, however, deals with the power outlets normally found in residential electrical wiring systems.

When viewing an electrical drawing, outlets are indicated by symbols (usually a small circle with appropriate markings to indicate the type of outlet). The most common symbols for receptacles are shown in *Figure 35*.

7.4.1 Branch Circuit Drawings

In the past, with the exception of very large residences and tract-development houses, the size of the average residential electrical system was not large enough to justify the expense of preparing complete electrical working drawings and specifications. Such electrical systems were either laid out by the architect in the form of a sketchy outlet arrangement, or laid out by the electrician on the job as the work progressed. However, many technical developments in residential electrical use—such as electric heat with sophisticated control wiring, increased use of electrical appliances, various electronic alarm systems, new lighting techniques, and the need for energy conservation techniques—have greatly expanded the demand and extended the complexity of today's residential electrical systems.

Each year, the number of homes with electrical systems designed by consulting engineering firms increases. Such homes are provided with complete electrical working drawings and specifications, similar to those frequently provided for commercial and industrial projects. Still, these are more the exception than the rule. Most residential projects will not have a complete set of drawings.

Circuit layout is provided on the drawings to follow for several reasons:

- They provide a visual layout of house wiring circuitry.
- They provide a sample of electrical residential drawings that are prepared by consulting engineering firms, although the number may still be limited.
- They introduce the method of showing electrical systems on working drawings to provide a foundation for tackling advanced electrical systems.

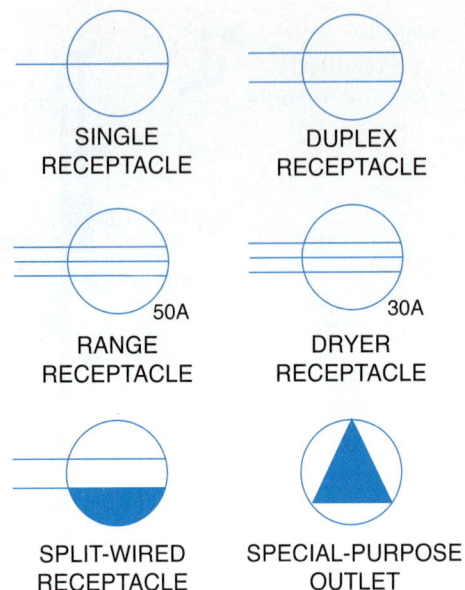

110F35.EPS

Figure 35   Typical outlet symbols appearing in electrical drawings.

Branch circuits are shown on electrical drawings by means of a single line drawn from the panelboard (or by homerun arrowheads indicating that the circuit goes to the panelboard) to the outlet, or from outlet to outlet where there is more than one outlet on the circuit.

The lines indicating branch circuits can be solid to show that the conductors are to be run concealed in the ceiling or wall; dashed to show that the conductors are to be run in the floor or ceiling below; or dotted to show that the wiring is to be run exposed. *Figure 36* shows examples of these three types of branch circuit lines.

In *Figure 36*, No. 12 indicates the wire size. The slash marks shown through the circuits in *Figure 36* indicate the number of current-carrying conductors in the circuit. Although two slash marks are shown for the current-carrying conductors (along with one slash mark for the ground), in actual practice, a branch circuit containing only two conductors usually contains no slash marks; that is, any circuit with no slash marks is assumed to have two conductors. However, three or more conductors are always indicated on electrical working drawings—either by slash marks for each conductor, or else by a note.

Never assume that you know the meaning of any electrical symbol. Although great efforts have been made in recent years to standardize drawing symbols, architects, consulting engineers, and

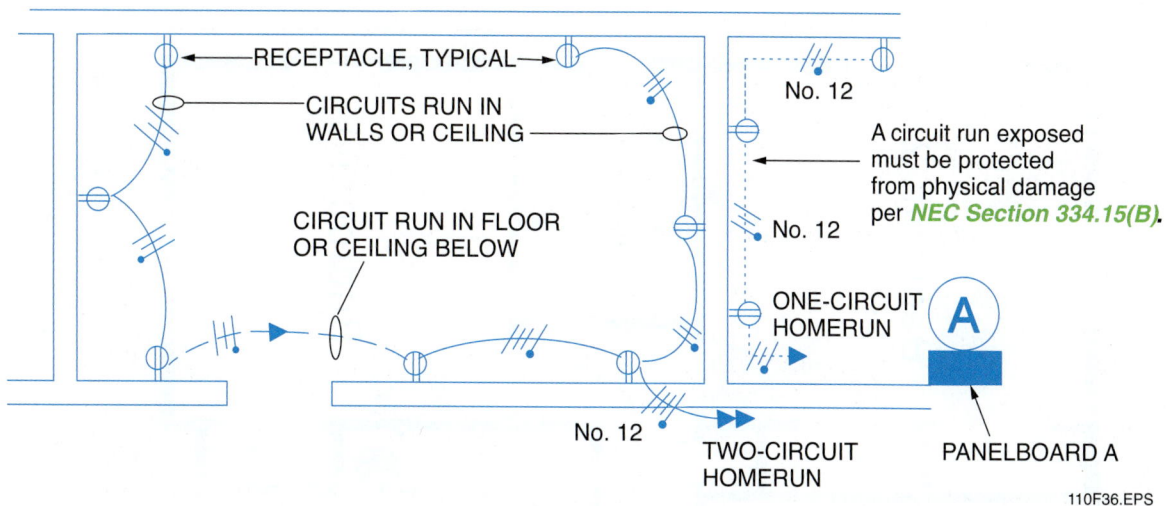

Diagram labels:
RECEPTACLE, TYPICAL
CIRCUITS RUN IN WALLS OR CEILING
CIRCUIT RUN IN FLOOR OR CEILING BELOW
No. 12
A circuit run exposed must be protected from physical damage per *NEC Section 334.15(B)*.
No. 12
ONE-CIRCUIT HOMERUN
A
No. 12
TWO-CIRCUIT HOMERUN
PANELBOARD A
110F36.EPS

Figure 36 Types of branch circuit lines shown on electrical working drawings.

electrical drafters still modify existing symbols or devise new ones to meet their own needs. Always consult the symbol list or legend on electrical working drawings for an exact interpretation of the symbols used.

7.4.2 Locating Receptacles

NEC Section 210.52(A) states the minimum requirements for the location of receptacles in dwelling units. It specifies that in each kitchen, family room, and dining room, receptacle outlets shall be installed so that no point along the floor line in any wall space is more than 6', measured horizontally, from an outlet in that space, including any wall space 2' or more in width and the wall space occupied by fixed panels in exterior walls, but excluding sliding panels. This means that the outlets will be no more than 12' apart. When spaced in this manner, a 6' extension cord will reach a receptacle at any point along the wall line. Receptacle outlets shall, insofar as practicable, be spaced equal distances apart. Receptacle outlets in floors shall not be counted as part of the required number of receptacle outlets unless located within 18" of the wall.

The *NEC*® defines wall space as a wall that is unbroken along the floor line by doorways, fireplaces, or similar openings. Each wall space that is two feet or more in width must be treated individually and separately from other wall spaces within the room.

The purpose of *NEC Section 210.52(A)* is to minimize the use of cords across doorways, fireplaces, and similar openings.

Figure 37 shows the outlets for a sample residence. In laying out these receptacle outlets, the floor line of the wall is measured (also around corners), but not across doorways, fireplaces, passageways, or other spaces where a flexible cord extended across the space would be unsuitable.

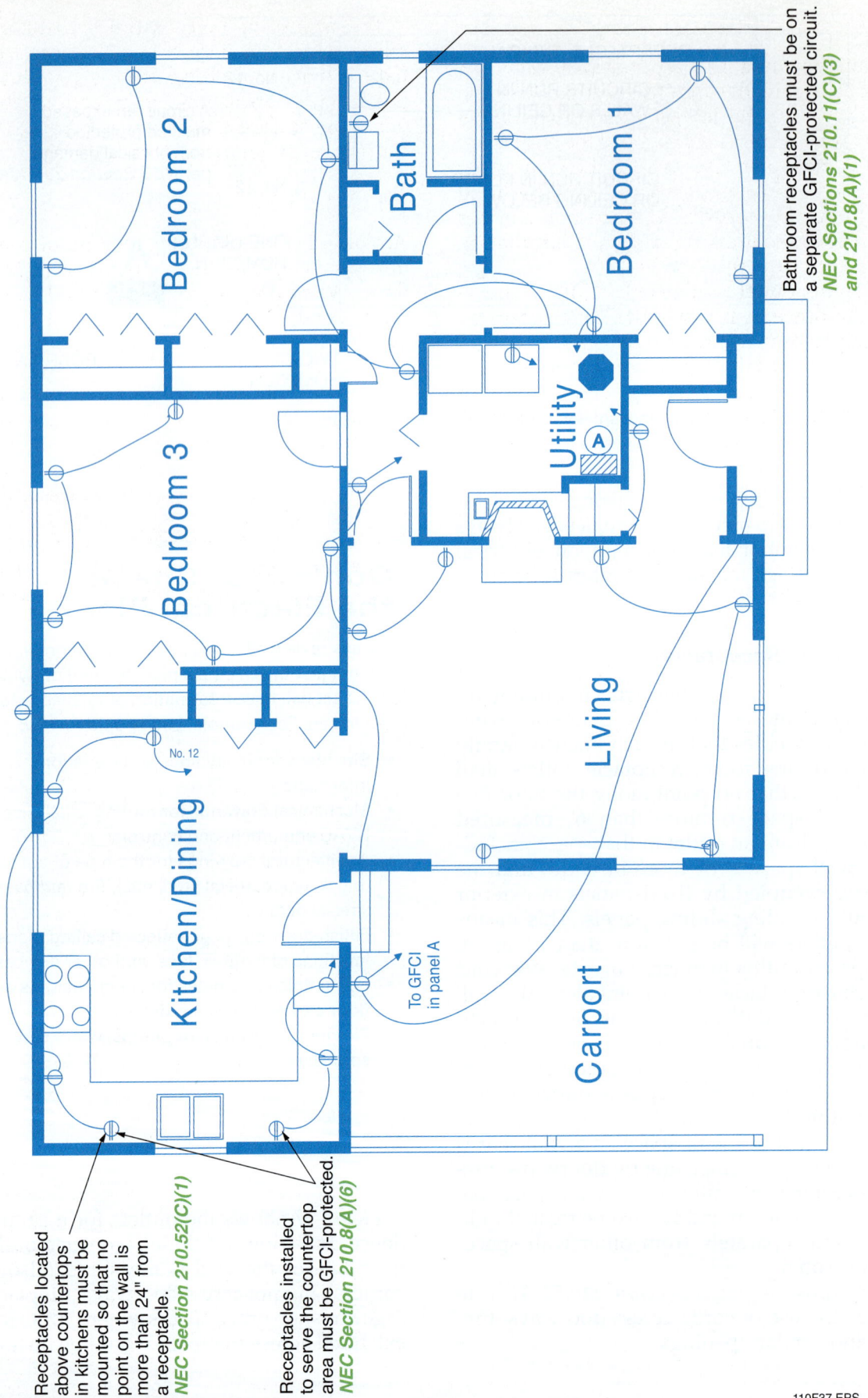

Bathroom receptacles must be on a separate GFCI-protected circuit. *NEC Sections 210.11(C)(3) and 210.8(A)(1)*

Bedroom 2

Bath

Bedroom 1

Bedroom 3

Utility

A

No. 12

Kitchen/Dining

Living

To GFCI in panel A

Carport

Receptacles located above countertops in kitchens must be mounted so that no point on the wall is more than 24" from a receptacle. *NEC Section 210.52(C)(1)*

Receptacles installed to serve the countertop area must be GFCI-protected. *NEC Section 210.8(A)(6)*

110F37.EPS

Figure 37   Floor plan of a sample residence.

8.0.0 LIGHTING FLOOR PLAN

A skeleton view of a lighting floor plan is shown in *Figure 38*. Again, the architect's/engineer's title blocks appear in the lower right corner of the drawing. A key plan appears above the engineer's title block. This plan is drawn to the same scale as the power plan; that is, ⅛" = 1'–0". A lighting fixture (luminaire) schedule appears in the upper right corner of the drawing and some installation notes appear below the schedule.

The lighting outlet symbols found on the drawing for the office/warehouse building represent both incandescent and fluorescent types; a circle on most electrical drawings usually represents an incandescent fixture, and a rectangle represents a fluorescent one. All of these symbols are designed to indicate the physical shape of a particular fixture and are usually drawn to scale.

The type of mounting used for all lighting fixtures is usually indicated in a lighting fixture schedule, which in this case is shown on the drawings. On some projects, the schedule may be found only in the written specifications.

The type of lighting fixture is identified by a numeral placed inside a triangle near each lighting fixture. If one type of fixture is used exclusively in one room or area, the triangular indicator need only appear once with the word ALL lettered at the bottom of the triangle.

8.1.0 Drawing Schedules

A schedule is a systematic method of presenting notes or lists of equipment on a drawing in tabular form. When properly organized and thoroughly understood, schedules are powerful timesaving devices for both those preparing the drawings and workers on the job.

For example, the lighting fixture schedule shown in *Figure 39* lists the fixture and identifies each fixture type on the drawing by number. The manufacturer and catalog number of each type are given along with the number, size, and type of lamp for each.

At times, all of the same information found in schedules will be duplicated in the written specifications, but combing through page after page

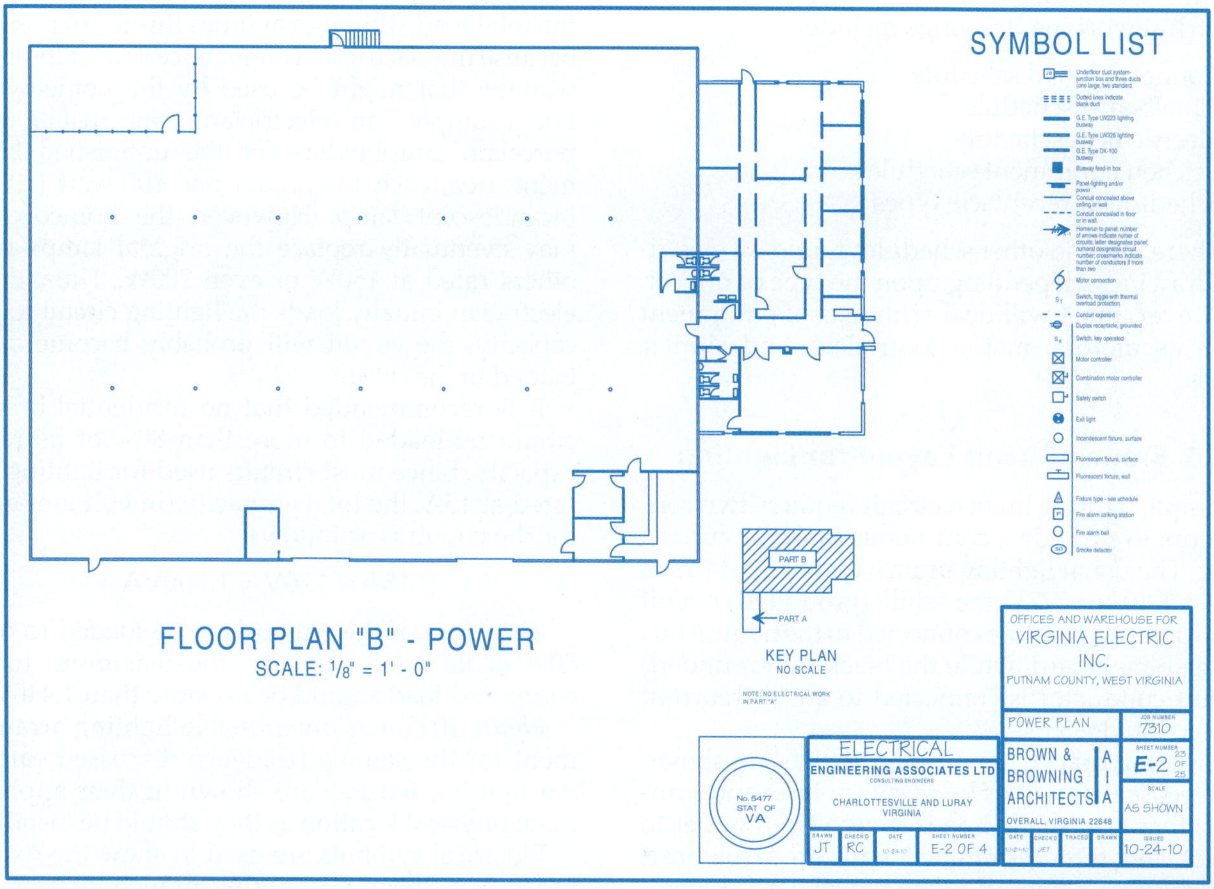

Figure 38 Sample lighting plan.

LIGHTING FIXTURE SCHEDULE

SYMBOL	TYPE	MANUFACTURER AND CATALOG NUMBER	MOUNTING	LAMPS
⊤▭	A	LIGHTOLIER 10234	WALL	2-40W T-12WWX
▭	B	LIGHTOLIER 10420	SURFACE	2-40W T-12 WWX
⊗	C	ALKCO RPC-210-6E	SURFACE	2-8W T-5
⊢○	D	P 7 S AL 2936	WALL	1-100W 'A'
○	E	P 7 S 110	SURFACE	1-100W 'A'

110F39.EPS

Figure 39   Lighting fixture (luminaire) schedule.

of written specifications can be time consuming. Workers do not always have access to the specifications while on the job, whereas they usually do have access to the working drawings. Therefore, the schedule is an excellent means of providing essential information in a clear and accurate manner, allowing the workers to carry out their assignments in the least amount of time.

Other schedules that are frequently found on electrical working drawings include:

- Connected load schedule
- Panelboard schedule
- Electric heat schedule
- Kitchen equipment schedule
- Schedule of receptacle types

There are also other schedules found on electrical drawings, depending upon the type of project. However, most will deal with lists of equipment such as motors, motor controllers, and similar items.

8.2.0 Branch Circuit Layout for Lighting

A simple lighting branch circuit requires two conductors to provide a continuous path for current flow. The usual lighting branch circuit operates at either 120V or 277V; the white (grounded) circuit conductor is therefore connected to the neutral bus in the panelboard, while the black (ungrounded) circuit conductor is connected to an overcurrent protection device.

Lighting branch circuits and outlets are shown on electrical drawings by means of lines and symbols; that is, a single line is drawn from outlet to outlet and then terminated with an arrowhead to indicate a homerun to the panelboard. Several methods are used to indicate the number and size of conductors, but the most common is to indicate the number of conductors in the circuit by using slash marks through the circuit lines and then indicate the wire size by a notation adjacent to these slash marks.

The circuits used to feed residential lighting must conform to standards established by the *NEC*® as well as by local and state ordinances. Most of the lighting circuits should be calculated to include the total load, although at times this is not possible because the electrician cannot be certain of the exact wattage that might be used by the homeowner. For example, an electrician may install four porcelain lampholders for the unfinished basement area, each to contain one 100-watt (100W) incandescent lamp. However, the homeowners may eventually replace the original lamps with others rated at 150W or even 200W. Thus, if the electrician initially loads the lighting circuit to full capacity, the circuit will probably become overloaded in the future.

It is recommended that no residential branch circuit be loaded to more than 80% of its rated capacity. Since most circuits used for lighting are rated at 15A, the total ampacity (in volt-amperes) for the circuit is as follows:

$$15A \times 120V = 1,800VA$$

Therefore, if the circuit is to be loaded to only 80% of its rated capacity, the maximum initial connected load should be no more than 1,440VA.

Figure 40 shows one possible lighting arrangement for the sample residence discussed earlier. All lighting fixtures are shown in their approximate physical location as they should be installed.

Electrical symbols are used to show the fixture types. Switches and lighting branch circuits are also shown by appropriate lines and symbols. The

NCCER — *Contren® Learning Series* 26110-11

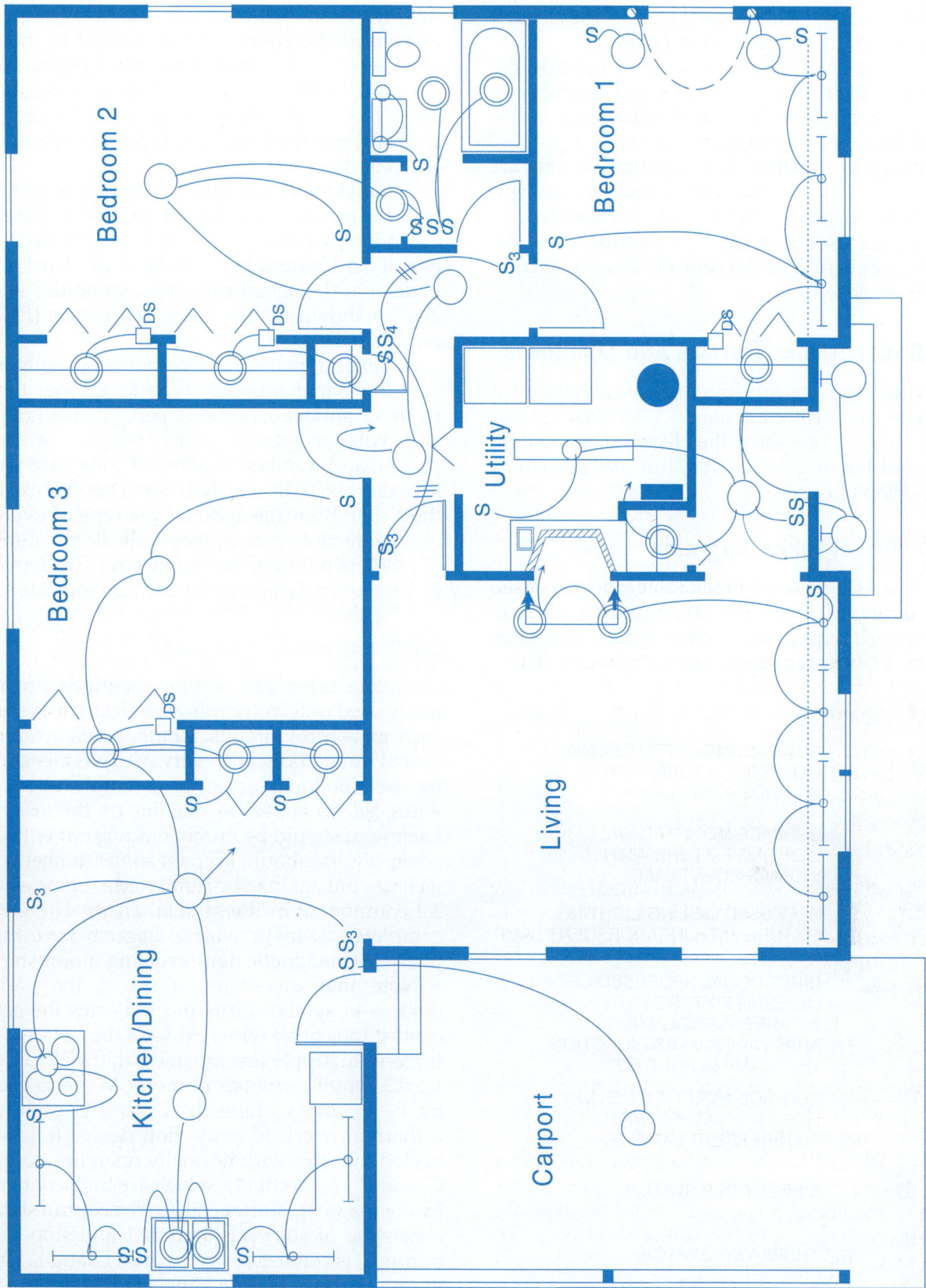

Figure 40   Lighting layout of the sample residence.

110F40.EPS

meanings of the symbols used on this drawing are explained in the symbol list in *Figure 41*.

In actual practice, the location of lighting fixtures and their related switches will probably be the extent of the information shown on working drawings. The circuits shown in *Figure 40* are meant to illustrate how lighting circuits are routed, not to imply that such drawings are typical for residential construction. If incandescent fixtures are used in a closet, they must meet the requirements of *NEC Section 410.16* and be completely enclosed.

9.0.0 ELECTRICAL DETAILS AND DIAGRAMS

Electrical diagrams are drawings that are intended to show electrical components and their related connections. They show the electrical association of the different components, but are seldom, if ever, drawn to scale.

9.1.0 Power-Riser Diagrams

One-line (single-line) block diagrams are used extensively to show the arrangement of electric service equipment. Power-riser diagrams (*Figure 42*) are typical of such drawings. These drawings show all pieces of electrical equipment as well as the connecting lines used to indicate service-entrance conductors and feeders. Notes are used to identify the equipment, indicate the size of conduit necessary for each feeder, and show the number, size, and type of conductors in each conduit.

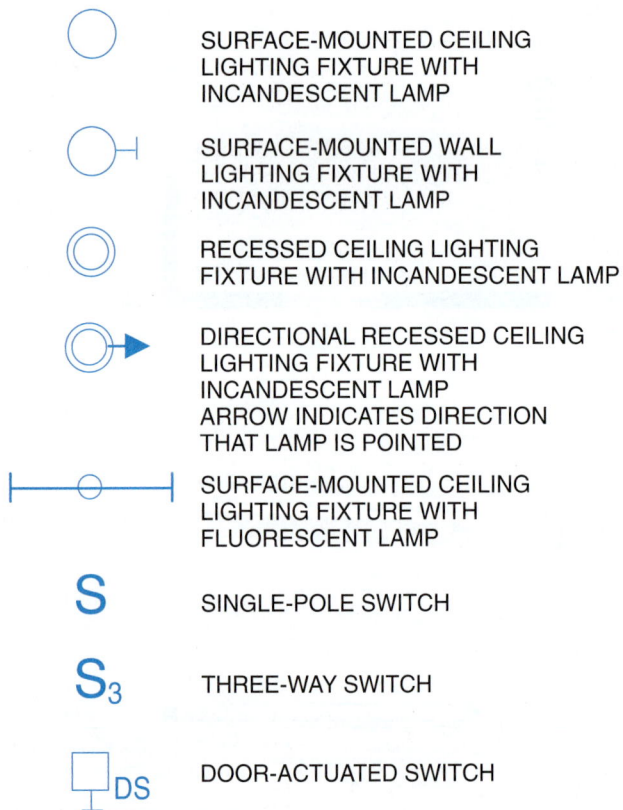

A panelboard schedule (*Figure 43*) is included with the power-riser diagram to indicate the exact components contained in each panelboard. This panelboard schedule is for the main distribution panel. On the actual drawings, schedules would also be shown for the other two panels (PNL A and PNL B).

In general, panelboard schedules usually indicate the panel number, type of cabinet (either flush- or surface-mounted), panel mains (ampere and voltage rating), phase (single- or three-phase), and number of wires. A four-wire panel, for example, indicates that a solid neutral exists in the panel. Branches indicate the type of overcurrent protection; that is, they indicate the number of poles, trip rating, and frame size. The items fed by each overcurrent device are also indicated.

9.2.0 Schematic Diagrams

Complete schematic wiring diagrams are normally used only in complicated electrical systems, such as control circuits. Components are represented by symbols, and every wire is either shown by itself or included in an assembly of several wires, which appear as one line on the drawing. Each wire should be numbered when it enters an assembly and should keep the same number when it comes out again to be connected to some electrical component in the system. *Figure 44* shows a complete schematic wiring diagram for a three-phase, AC magnetic non-reversing motor starter.

Note that this diagram shows the various devices in symbol form and indicates the actual connections of all wires between the devices. The three-wire supply lines are indicated by L_1, L_2, and L_3; the motor terminals of motor M are indicated by T_1, T_2, and T_3. Lines L_1, L_2, and L_3 each have a thermal overload protection device (OL) connected in series with normally open line contacts C_1 and C_3, respectively, which are both controlled by the magnetic starter coil, C. The control station, consisting of start pushbutton 1 and stop pushbutton 2, is connected across lines L_1 and L_2. Auxiliary contacts (C_4) are connected in series with the stop pushbutton and in parallel with the start pushbutton. The control circuit also has normally closed overload contacts (OC) connected in series with the magnetic starter coil (C).

⭕	SURFACE-MOUNTED CEILING LIGHTING FIXTURE WITH INCANDESCENT LAMP
⭕─	SURFACE-MOUNTED WALL LIGHTING FIXTURE WITH INCANDESCENT LAMP
◎	RECESSED CEILING LIGHTING FIXTURE WITH INCANDESCENT LAMP
◎▶	DIRECTIONAL RECESSED CEILING LIGHTING FIXTURE WITH INCANDESCENT LAMP ARROW INDICATES DIRECTION THAT LAMP IS POINTED
├─○─┤	SURFACE-MOUNTED CEILING LIGHTING FIXTURE WITH FLUORESCENT LAMP
S	SINGLE-POLE SWITCH
S_3	THREE-WAY SWITCH
DS	DOOR-ACTUATED SWITCH

110F41.EPS

Figure 41 Electrical symbols list.

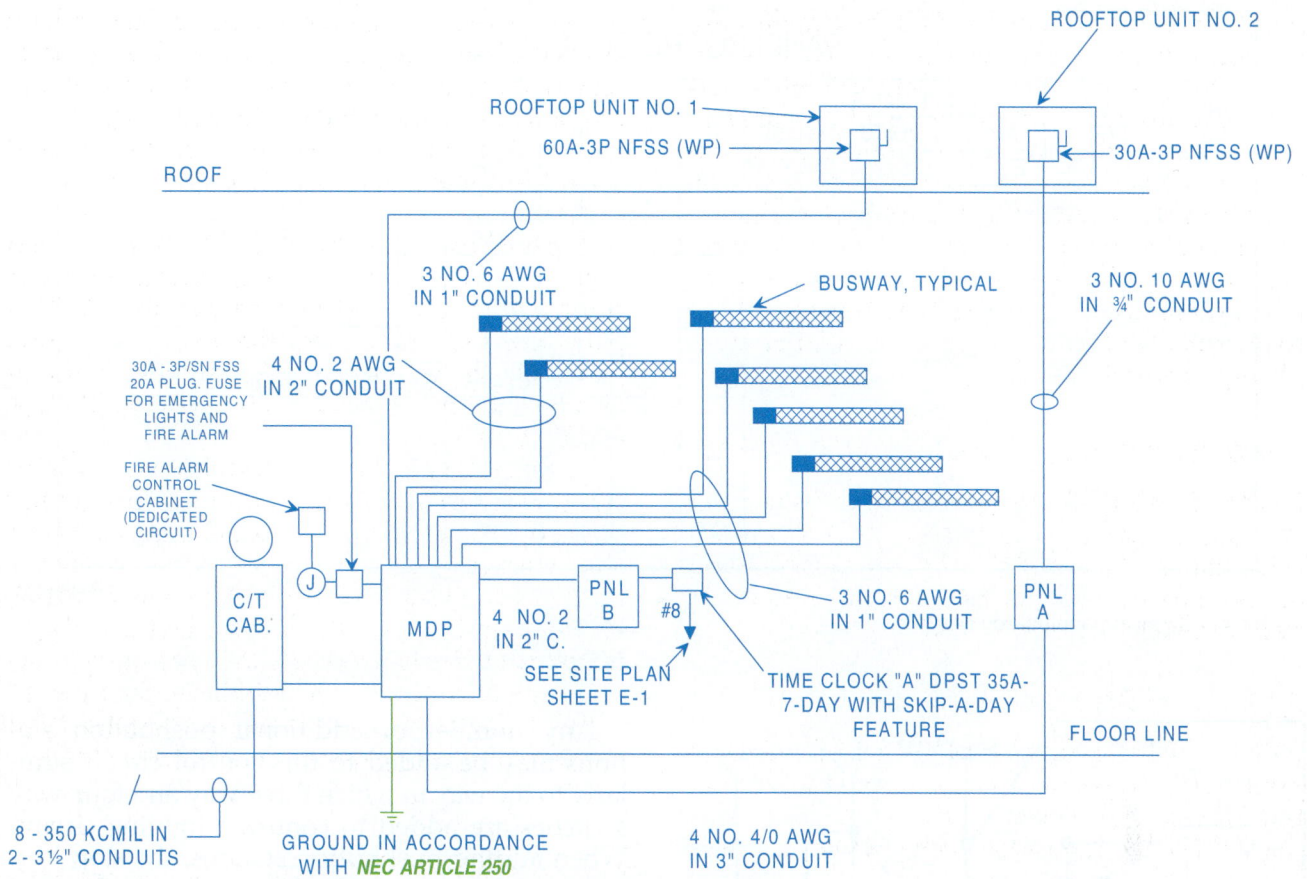

COMMERCIAL

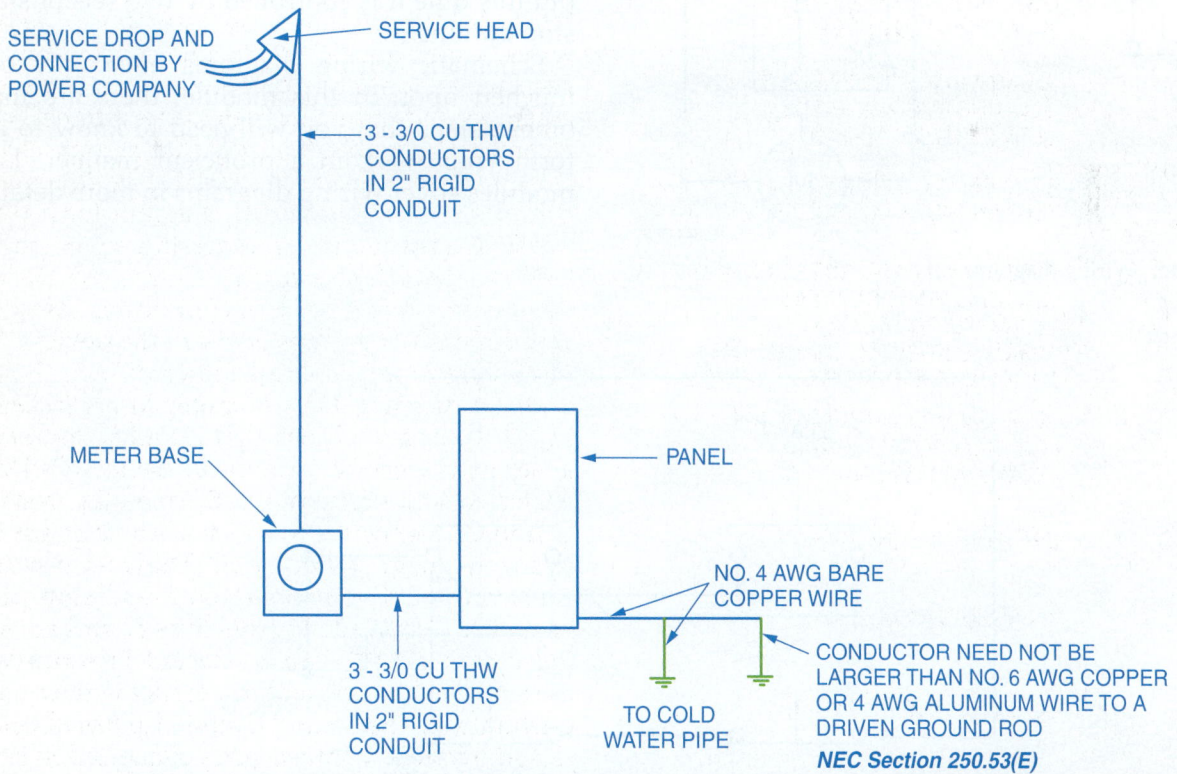

RESIDENTIAL

110F42.EPS

Figure 42 Typical power-riser diagrams.

PANELBOARD SCHEDULE

PANEL No.	CABINET TYPE	PANEL MAINS			BRANCHES					ITEMS FED OR REMARKS
		AMPS	VOLTS	PHASE	1P	2P	3P	PROT.	FRAME	
MDP	SURFACE	600A	120/208	3φ,4-W	-	-	1	225A	25,000	PANEL "A"
					-	-	1	100A	18,000	PANEL "B"
					-	-	1	100A		POWER BUSWAY
					-	-	1	60A		LIGHTING BUSWAY
					-	-	1	70A		ROOFTOP UNIT #1
					-	-	1	70A	▼	SPARE
					-	-	1	600A	42,000	MAIN CIRCUIT BRKR

110F43.EPS

Figure 43   Typical panelboard schedule.

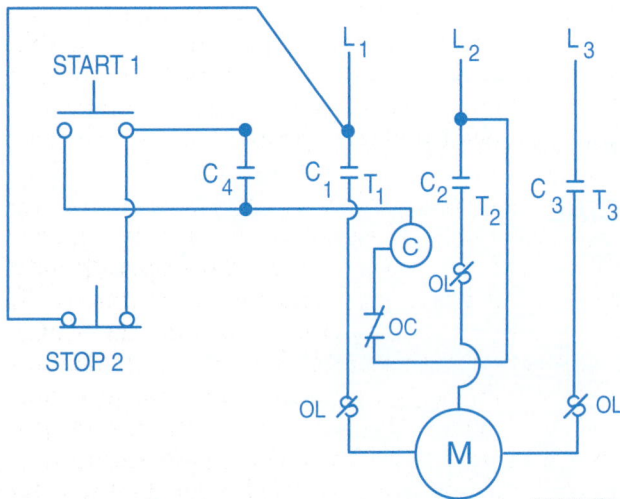

110F44.EPS

Figure 44   Wiring diagram.

Any number of additional pushbutton stations may be added to this control circuit similarly to the way in which three-way and four-way switches are added to control a lighting circuit. When adding pushbutton stations, the stop buttons are always connected in series and the start buttons are always connected in parallel. *Figure 45* shows the same motor starter circuit in *Figure 44*, but this time it is controlled by two sets of start/stop buttons.

Schematic wiring diagrams have only been touched upon in this module; there are many other details that you will need to know to perform your work in a proficient manner. Later modules cover wiring diagrams in more detail.

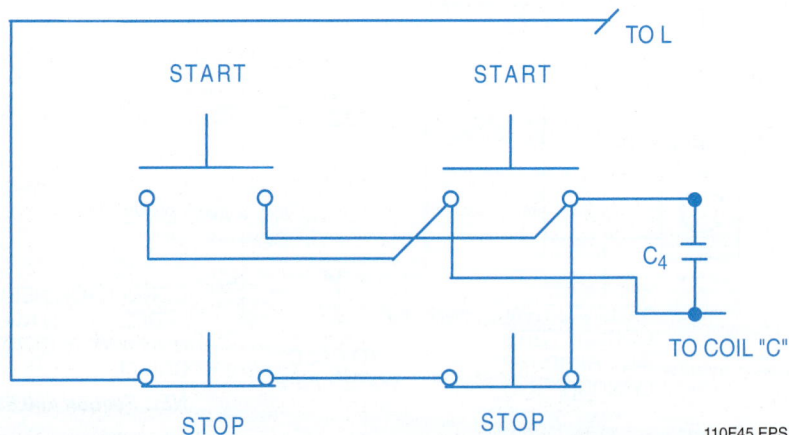

110F45.EPS

Figure 45   Circuit being controlled by two sets of start/stop buttons.

9.3.0 Drawing Details

A detail drawing is a drawing of a separate item or portion of an electrical system, giving a complete and exact description of its use and all the details needed to show the electrician exactly what is required for its installation. For example, the power plan for the office/warehouse has a sectional cut through the busduct. This is a good example of where an extra, detailed drawing is desirable.

A set of electrical drawings will sometimes require large-scale drawings of certain areas that are not indicated with sufficient clarity on the small-scale drawings. For example, the site plan may show exterior pole-mounted lighting fixtures that are to be installed by the contractor.

10.0.0 WRITTEN SPECIFICATIONS

The written specifications for a building or project are the written descriptions of work and duties required of the owner, architect, and consulting engineer. Together with the working drawings, these specifications form the basis of the contract requirements for the construction of the building or project. Those who use the construction drawings and specifications must always be alert to discrepancies between the working drawings and the written specifications. These are some situations where discrepancies may occur:

- Architects or engineers use standard or prototype specifications and attempt to apply them without any modification to specific working drawings.
- Previously prepared standard drawings are changed or amended by reference in the specifications only and the drawings themselves are not changed.
- Items are duplicated in both the drawings and specifications, but an item is subsequently amended in one and overlooked in the other contract document.

In such instances, the person in charge of the project has the responsibility to ascertain whether the drawings or the specifications take precedence. Such questions must be resolved, preferably before the work begins, to avoid added costs to the owner, architect/engineer, or contractor.

10.1.0 How Specifications Are Written

Writing accurate and complete specifications for building construction is a serious responsibility for those who design the buildings because the specifications, combined with the working drawings,

govern practically all important decisions that are made during the construction span of every project. Compiling and writing these specifications is not a simple task, even for those who have had considerable experience in preparing such documents. A set of written specifications for a single project will usually contain thousands of products, parts, and components, and the methods of installing them, all of which must be covered in either the drawings and/or specifications. No one can memorize all of the necessary items required to describe accurately the various areas of construction. One must rely upon reference materials such as manufacturer's data, catalogs, checklists, and, most of all, a high-quality master specification.

10.2.0 Format of Specifications

For convenience in writing, speed in estimating, and ease of reference, the most suitable organization of the specifications is a series of sections dealing with the construction requirements, products, and activities that is easily understandable by the different trades. Those people who use the specifications must be able to find all the information they need without spending too much time looking for it.

The most commonly used specification-writing format used in North America is the *Master-Format*®. This standard was developed jointly by the Construction Specifications Institute (CSI) and Construction Specifications Canada (CSC). For many years prior to 2004, the organization of construction specifications and suppliers catalogs was based on a standard with 16 sections, otherwise known as divisions. The divisions and their subsections were individually identified by a five-digit numbering system. The first two digits represented the division number and the next three individual numbers represented successively lower levels of breakdown. For example, the number 13213 represents division 13, subsection 2, sub-subsection 1, and sub-sub-subsection 3. In this older version of the standard, electrical systems, including any electronic or special electrical systems, were lumped together under Division 16 – *Electrical*. Today, specifications conforming to the 16 division format may still be in use.

In 2004, the *MasterFormat*® standard underwent a major change. What had been 16 divisions was expanded to four major groupings and 49 divisions with some divisions reserved for future expansion. *MasterFormat*® was again updated for 2010 (*Figure 46*). The first 14 divisions are essentially the same as the old format. Subjects under the old Division 15 – *Mechanical*

MasterFormat GROUPS, SUBGROUPS, AND DIVISIONS

PROCUREMENT AND CONTRACTING REQUIREMENTS GROUP

Division 00 – Procurement and Contracting
        Requirements
    Introductory Information
    Procurement Requirements
    Contracting Requirements

SPECIFICATIONS GROUP

GENERAL REQUIREMENTS SUBGROUP
Division 01 – General Requirements

FACILITY CONSTRUCTION SUBGROUP
Division 02 – Existing Conditions
Division 03 – Concrete
Division 04 – Masonry
Division 05 – Metals
Division 06 – Wood, Plastics, and Composites
Division 07 – Thermal and Moisture Protection
Division 08 – Openings
Division 09 – Finishes
Division 10 – Specialties
Division 11 – Equipment
Division 12 – Furnishings
Division 13 – Special Construction
Division 14 – Conveying Equipment
Division 15 – Reserved for Future Expansion
Division 16 – Reserved for Future Expansion
Division 17 – Reserved for Future Expansion
Division 18 – Reserved for Future Expansion
Division 19 – Reserved for Future Expansion

FACILITY SERVICES SUBGROUP
Division 20 – Reserved for Future Expansion
Division 21 – Fire Suppression

Division 22 – Plumbing
Division 23 – Heating, Ventilating, and Air-
        Conditioning (HVAC)
Division 24 – Reserved for Future Expansion
Division 25 – Integrated Automation
Division 26 – Electrical
Division 27 – Communications
Division 28 – Electronic Safety and Security
Division 29 – Reserved for Future Expansion

SITE AND INFRASTRUCTURE SUBGROUP
Division 30 – Reserved for Future Expansion
Division 31 – Earthwork
Division 32 – Exterior Improvements
Division 33 – Utilities
Division 34 – Transportation
Division 35 – Waterway and Marine
        Construction
Division 36 – Reserved for Future Expansion
Division 37 – Reserved for Future Expansion
Division 38 – Reserved for Future Expansion
Division 39 – Reserved for Future Expansion

PROCESS EQUIPMENT SUBGROUP
Division 40 – Process Integration
Division 41 – Material Processing and Handling
        Equipment
Division 42 – Process Heating, Cooling, and
        Drying Equipment
Division 43 – Process Gas and Liquid Handling,
        Purification, and Storage
        Equipment
Division 44 – Pollution and Waste Control
        Equipment
Division 45 – Industry-Specific Manufacturing
        Equipment
Division 46 – Water and Wastewater Equipment
Division 47 – Reserved for Future Expansion
Division 48 – Electrical Power Generation
Division 49 – Reserved for Future Expansion

110F46.EPS

Figure 46   2010 *MasterFormat*®.

Specifications

Written specifications supplement the related working drawings in that they contain details not shown on the drawings. Specifications define and clarify the scope of the job. They describe the specific types and characteristics of the components that are to be used on the job and the methods for installing some of them. Many components are identified specifically by the manufacturer's model and part numbers. This type of information is used to purchase the various items of hardware needed to accomplish the installation in accordance with the contractual requirements.

have been relocated to new divisions 22 and 23. The basic subjects under old Division 16 – *Electrical* have been relocated to new divisions 26 and 27. In addition, the numbering system was changed to 6 digits to allow for more subsections in each division, which allowed for finer definition. In the new numbering system, the first two digits represent the division number. The next two digits represent subsections of the division and the two remaining digits represent the third level sub-subsection numbers. The fourth level, if required, is a decimal and number added to the end of the last two digits. For example, the number 132013.04 represents division 13, subsection 20, sub-subsection 13, and sub-sub-subsection 04. Under the new standard, the Facility Service Subgroup contains the divisions that are most important to the electrician. These include the following divisions:

- *Division 25 – Integrated Automation*
- *Division 26 – Electrical*
- *Division 27 – Communications*
- *Division 28 – Electronic Safety and Security*

Figure 47 contains a detailed breakdown of the electrical division.

NUMBER	TITLE	EXPLANATION

DIVISION 26 – ELECTRICAL

26 00 00   Electrical

may be used as division level section title.

See: 02 41 19 for selective demolition of existing electrical systems.
03 30 00 for cast-in-place concrete equipment bases.
07 84 00 for firestopping.
07 92 00 for joint sealants.
08 31 00 for access doors and panels.
09 91 00 for field painting.
31 23 33 for trenching and backfilling.

---

26 01 00   Operation and Maintenance of Electrical Systems

Includes: maintenance, repair, rehabilitation, replacement, restoration, preservation, etc. of electrical systems. medium voltage: 2400 V to 69 kV. low voltage: 600 V and less.

Notes: Definitions medium voltage: 2400 V to 69 kV. low voltage: 600 V and less.

Level 4 Numbering Recommendation: following numbering is recommended for the creation of Level 4 titles:
.51-.59 for maintenance.
.61-.69 for repair.
.71-.79 for rehabilitation.
.81-.89 for replacement.
.91-.99 for restoration.

26 01 10	Operation and Maintenance of Medium-Voltage Electrical Distribution
26 01 20	Operation and Maintenance of Low-Voltage Electrical Distribution
26 01 26	Maintenance Testing of Electrical Systems
26 01 30	Operation and Maintenance of Facility Electrical Power Generating and Storing Equipment
26 01 40	Operation and Maintenance of Electrical and Cathodic Protection Systems
26 01 50	Operation and Maintenance of Lighting
26 01 50.51	Luminaire Relamping
26 01 50.81	Luminaire Replacement

26 05 00   Common Work Results for Electrical

Includes: subjects common to multiple titles in Division 26. raceway and boxes includes conduit, tubing, surface raceways, and electrical boxes. medium voltage: 2400 V to 69 kV. low voltage: 600 V and less. control voltage: 50 V

311

110F47A.EPS

Figure 47   Detailed breakdown of the electrical division (1 of 13).

and less.

Alternate Terms/Abbreviations: EMT: electrical metallic tubing.

Notes: Definitions medium voltage: 2400 V to 69 kV. low voltage: 600 V and less. control voltage: 50 V and less.

See 01 80 00 for performance requirements of subjects common to multiple titles.
05 35 00 for raceway decking assemblies.
05 45 16 for electrical metal supports.
13 48 00 for sound, vibration, and seismic control.
25 05 13 for conductors and cables for integrated automation.
25 05 26 for grounding and bonding for integrated automation.
25 05 28 for pathways for integrated automation.
25 05 48 for vibration and seismic control for integrated automation.
25 05 53 for identification for integrated automation.
27 05 28 for pathways for communications systems.
27 05 46 for utility poles for communications systems.
27 05 48 for vibration and seismic controls for communications.
27 05 53 for identification for communications.
28 05 13 for conductors and cables for electronic safety and security.
28 05 26 for grounding and bonding for electronic safety and security.
28 05 28 for pathways for electronic safety and security.
28 05 48 for vibration and seismic controls for electronic safety and security.
28 05 53 for identification for electronic safety and security.
33 71 16 for electrical utility poles.
33 71 19 for electrical utility underground ducts and manholes.

26 05 13	Medium-Voltage Cables	
	26 05 13.13	Medium-Voltage Open Conductors
	26 05 13.16	Medium-Voltage, Single- and Multi-Conductor Cables
26 05 19	Low-Voltage Electrical Power Conductors and Cables	
	26 05 19.13	Undercarpet Electrical Power Cables
	26 05 19.23	Manufactured Wiring Assemblies
26 05 23	Control-Voltage Electrical Power Cables	
26 05 26	Grounding and Bonding for Electrical Systems	
26 05 29	Hangers and Supports for Electrical Systems	
26 05 33	Raceway and Boxes for Electrical Systems	
	26 05 33.13	Conduit for Electrical Systems
	26 05 33.16	Boxes for Electrical Systems
	26 05 33.23	Surface raceways for Electrical Systems

312

110F47B.EPS

Figure 47 Detailed breakdown of the electrical division (2 of 13).

NUMBER	TITLE	EXPLANATION
26 05 36	Cable Trays for Electrical Systems	
26 05 39	Underfloor Raceways for Electrical Systems	
26 05 43	Underground Ducts and Raceways for Electrical Systems	
26 05 46	Utility Poles for Electrical Systems	
26 05 48	Vibration and Seismic Controls for Electrical Systems	
26 05 53	Identification for Electrical Systems	
26 05 73	Overcurrent Protective Device Coordination Study	
26 05 83	Wiring Connections	

26 06 00 Schedules for Electrical

Notes: a schedule may be included on drawings, in the project manual, or a project book.

Definitions: medium voltage: 2400 V to 69 kV. low voltage: 600 V and less.

Includes: schedules of items common to multiple titles in Division 26.

26 06 10	Schedules for Medium-Voltage Electrical Distribution	
26 06 20	Schedules for Low-Voltage Electrical Distribution	
	26 06 20.13	Electrical Switchboard Schedule
	26 06 20.16	Electrical Panelboard Schedule
	26 06 20.19	Electrical Motor-Control Center Schedule
	26 06 20.23	Electrical Circuit Schedule
	26 06 20.26	Wiring Device Schedule
26 06 30	Schedules for Facility Electrical Power Generating and Storing Equipment	
26 06 40	Schedules for Electrical and Cathodic Protection Systems	
26 06 50	Schedules for Lighting	
	26 06 50.13	Lighting Panelboard Schedule
	26 06 50.16	Lighting Fixture Schedule

26 08 00 Commissioning of Electrical Systems

Includes: commissioning of items common to multiple titles in Division 26.

See: 01 91 00 for commissioning of subjects common to multiple divisions.

26 09 00 Instrumentation and Control for Electrical Systems

Includes: instrumentation and control associated with electrical systems.

See: 13 50 00 for special instrumentation.
25 36 00 for integrated automation instrumentation and terminal devices for electrical systems.
25 56 00 for integrated automation control of electrical systems.

313

110F47C.EPS

Figure 47   Detailed breakdown of the electrical division (3 of 13).

25 96 00 for integrated automation control sequences for electrical systems.
33 09 70 for instrumentation and control for electrical utilities.

26 09 13	Electrical Power Monitoring
26 09 15	Peak Load Controllers
26 09 16	Electrical Controls and Relays
26 09 17	Programmable Controllers
26 09 19	Enclosed Contactors
26 09 23	Lighting Control Devices

Includes: clock and calendar, photoelectric switches, occupancy sensors, and light-leveling control devices. control- and low-voltage lighting control devices connected through computers. addressable lighting control devices and lighting components (ballasts) connected through computers.

See: 11 61 00 for theater and stage equipment.
26 50 00 for lighting.
26 55 61 for theatrical lighting.

See Also: 11 61 00 for theatrical lighting controls.

26 09 26	Lighting Control Panelboards
26 09 33	Central Dimming Controls
	26 09 33.13 Multichannel Remote-Controlled Dimmers
	26 09 33.16 Remote-Controlled Dimming Stations
26 09 36	Modular Dimming Controls
	26 09 36.13 Manual Modular Dimming Controls
	26 09 36.16 Integrated Multipreset Modular Dimming Controls
26 09 43	Network Lighting Controls
	26 09 43.13 Digital-Network Lighting Controls
	26 09 43.16 Addressable Fixture Lighting Control
26 09 61	Theatrical Lighting Controls

26 10 00 Medium-Voltage Electrical Distribution

Includes: substations, transformers, switchgear, and circuit protection devices to distribute medium-voltage electrical power from the facility service point to the point of delivery.

Notes: Definitions medium voltage: 2400 V to 69 kV.

See 26 05 13 for medium-voltage cables.
26 20 00 for low-voltage electrical distribution.
26 30 00 for facility electrical power generating and storing equipment.
33 71 00 for electrical utility distribution.

314

110F47D.EPS

Figure 47   Detailed breakdown of the electrical division (4 of 13).

NUMBER	TITLE	EXPLANATION

26 11 00   Substations

Includes: assembly of switches, circuit breakers, buses, and transformers to switch circuits and convert power from one voltage to another.

See: 33 72 00 for utility substations.
34 21 16 for traction power substations.

26 11 13   Primary Unit Substations
26 11 16   Secondary Unit Substations

26 12 00   Medium-Voltage Transformers

Includes: transformers for medium-voltage applications.

See: 26 22 00 for low-voltage transformers.
33 73 00 for utility transformers.
34 21 23 for traction power transformer-rectifier units.

26 12 13   Liquid-Filled, Medium-Voltage Transformers
26 12 16   Dry-Type, Medium-Voltage Transformers
26 12 19   Pad-Mounted, Liquid-Filled, Medium-Voltage Transformers

26 13 00   Medium-Voltage Switchgear

Includes: switchgear for medium-voltage applications.

See: 26 23 00 for low-voltage switchgear.
33 77 00 for medium-voltage utility switchgear.
34 21 19 for traction power switchgear.

26 13 13   Medium-Voltage Circuit Breaker Switchgear
26 13 16   Medium-Voltage Fusible Interrupter Switchgear
26 13 19   Medium-Voltage Vacuum Interrupter Switchgear
26 13 23   Medium-Voltage Metal-Enclosed Switchgear
26 13 26   Medium-Voltage Metal-Clad Switchgear
26 13 29   Medium-Voltage Compartmentalized Switchgear

26 16 00   Medium-Voltage Metering

26 18 00   Medium-Voltage Circuit Protection Devices

Includes: circuit protection devices for medium-voltage applications.

See: 26 28 00 for low-voltage circuit protective devices.
26 41 23 for lightning protection surge arresters and suppressors.
33 77 00 for medium-voltage utility circuit protection devices.

26 18 13   Medium-Voltage Cutouts

315

110F47E.EPS

Figure 47   Detailed breakdown of the electrical division (5 of 13).

NUMBER	TITLE	EXPLANATION
26 18 16	Medium-Voltage Fuses	
26 18 19	Medium-Voltage Lightning Arresters	
26 18 23	Medium-Voltage Surge Arresters	
26 18 26	Medium-Voltage Reclosers	
26 18 29	Medium-Voltage Enclosed Bus	
26 18 33	Medium-Voltage Enclosed Fuse Cutouts	
26 18 36	Medium-Voltage Enclosed Fuses	
26 18 39	Medium-Voltage Motor Controllers	

26 20 00 Low-Voltage Electrical Transmission

Includes: overhead power systems, transformers, switchgear, switchboards, panelboards, enclosed bus assemblies, power distribution units, controllers, wiring devices, and circuit protection devices to distribute low-voltage electrical power from the point of voltage transformation to the point of use. typical voltages: 120, 208, 230, 240, 277, 460, and 480.

Notes: Definitions low voltage: 600 V and less.

See 26 05 19 for low-voltage electrical power conductors and cables.
26 10 00 for medium-voltage electrical distribution.
26 30 00 for facility electrical power generating and storing equipment.

26 21 00   Low-Voltage Electrical Service Entrance

See: 26 05 19 for low-voltage electrical power conductors and cables.
26 05 46 for utility poles for electrical systems.
33 71 13 for site electrical transmission towers.
33 71 16 for electrical utility poles.

| 26 21 13 | Low-Voltage Overhead Electrical Service Entrance |
| 26 21 16 | Low-Voltage Underground Electrical Service Entrance |

26 22 00   Low-Voltage Transformers

Includes: transformers for low-voltage applications.

See: 26 12 00 for medium-voltage transformers.

26 22 13	Low-Voltage Distribution Transformers
26 22 16	Low-Voltage Buck-Boost Transformers
26 22 19	Control and Signal Transformers

26 23 00   Low-Voltage Switchgear

Includes: switchgear for low-voltage applications.

316

110F47F.EPS

Figure 47   Detailed breakdown of the electrical division (6 of 13).

NUMBER	TITLE	EXPLANATION
		See: 26 13 00 for medium-voltage switchgear.
26 23 13	Paralleling Low-Voltage Switchgear	

26 24 00   Switchboards and Panelboards

Includes: switchboards, panelboards, and control centers.

See: 26 27 16 for electrical cabinets and enclosures.
26 29 13 for enclosed controllers.
26 29 23 for variable-frequency motor controllers.

26 24 13	Switchboards
26 24 16	Panelboards
26 24 19	Motor-Control Centers

26 25 00   Enclosed Bus Assemblies

Includes: busway, step bus, and tap boxes.

See: 33 72 26 for utility substation bus assemblies.

26 26 00   Power Distribution Units

Includes: distribution units with integral transformers, panelboards, and power conditioning components.

See: 26 24 16 for panelboards.

26 27 00   Low-Voltage Distribution Equipment

Includes: wiring devices includes receptacles, switches, dimmers, and finish plates.

See: 26 24 00 for switchboards and panelboards.
33 71 73 for utility electric meters.

26 27 13	Electricity Metering
26 27 16	Electrical Cabinets and Enclosures
26 27 19	Multi-Outlet Assemblies
26 27 23	Indoor Service Poles
26 27 26	Wiring Devices
26 27 73	Door Chimes

26 28 00   Low-Voltage Circuit Protective Devices

Includes: circuit protection devices for low-voltage applications. enclosed switches and transfer switches.

See: 26 18 00 for medium-voltage circuit protection devices.

317

110F47G.EPS

Figure 47 Detailed breakdown of the electrical division (7 of 13).

26 28 13   Fuses
26 28 16   Enclosed Switches and Circuit Breakers
               26 28 16.13   Enclosed Circuit Breakers
               26 28 16.16   Enclosed Switches

26 29 00   Low-Voltage Controllers

Includes: contactors and motor controllers.

May Include: fuses.

Alternate Terms/Abbreviations: enclosed controllers: motor controllers.

See: 26 24 19 for motor-control centers.
26 28 13 for fuses.

26 29 13   Enclosed Controllers
               26 29 13.13   Across-the-Line Motor Controllers
               26 29 13.16   Reduced-Voltage Motor Controllers
26 29 23   Variable-Frequency Motor Controllers
26 29 33   Controllers for Fire Pump Drivers
               26 29 33.13   Full-Service Controllers for Fire Pump Electric-Motor Drivers
               26 29 33.16   Limited-Service Controllers for Fire Pump Electric-Motor Drivers
               26 29 33.19   Controllers for Fire Pump Diesel Engine Drivers

26 30 00 Facility Electrical Power Generating and Storing Equipment

Includes: equipment to generate and store electrical power for a single facility.

Notes: 48 10 00 for electrical power generation equipment.

26 31 00   Photovoltaic Collectors

Includes: solar cells to convert sunlight to electricity.

See: 07 31 00 for solar collector roof shingles.
22 33 30 for residential, collector-to-tank, solar-electric domestic water heaters.
23 56 00 for solar energy heating equipment.
42 12 23 for solar process heaters.
42 13 26 for industrial solar radiation heat exchangers.
48 14 00 for solar energy electrical power generation equipment.

26 32 00   Packaged Generator Assemblies

Includes: generators, frequency changers, and rotary converters and uninterruptible power units.

318

110F47H.EPS

Figure 47   Detailed breakdown of the electrical division (8 of 13).

NUMBER	TITLE	EXPLANATION
		See: 23 11 00 for facility fuel piping. 23 24 00 for internal-combustion engine piping. 48 11 00 for fossil fuel plant electrical power generation equipment. 48 13 00 for hydroelectric plant electrical power generation equipment. 48 15 00 for wind energy electrical power generation equipment.
26 32 13	Engine Generators 26 32 13.13 Diesel-Engine-Driven Generator Sets 26 32 13.16 Gas-Engine-Driven Generator Sets 26 32 13.26 Gas-Turbine Engine-Driven Generators	
		Alternate Terms/Abbreviations: microturbines
		See: 48 11 23 Fossil Fuel Electrical Power Plant Gas Turbines
26 32 16	Steam-Turbine Generators	
26 32 19	Hydro-Turbine Generators	
26 32 23	Wind Energy Equipment	
26 32 26	Frequency Changers	
26 32 29	Rotary Converters	
26 32 33	Rotary Uninterruptible Power Units	

26 33 00 Battery Equipment

Includes: batteries, battery racks, battery chargers, static power converters, uninterruptible power supplies, and accessories.

May Include: battery-operated emergency light fixtures.

See: 25 36 23 for integrated automation battery monitors.
26 31 00 for photovoltaic collectors.
33 72 33 for electrical utility substation.
48 17 13 for electrical power generation batteries.

See Also: 26 52 00 for emergency lighting incorporating batteries.

26 33 13	Batteries	
26 33 16	Battery Racks	
26 33 19	Battery Units	
26 33 23	Central Battery Equipment	
26 33 33	Static Power Converters	
26 33 43	Battery Chargers	
26 33 46	Battery Monitoring	

319

110F47I.EPS

Figure 47   Detailed breakdown of the electrical division (9 of 13).

NUMBER	TITLE	EXPLANATION
26 33 53	Static Uninterruptible Power Supply	

26 35 00 Power Filters and Conditioners

Includes: capacitors, chokes and inductors, filters, power factor controllers, and voltage regulators.

Alternate Terms/Abbreviations: EMI: electromagnetic interference. RFI: radio frequency interference. power factor correction equipment: power factor controllers.

See: 08 34 46 for RFI shielding doors.
08 56 46 for RFI shielding windows.
13 49 00 for radiation protection.
26 18 23 for medium-voltage surge arresters.
28 32 00 for radiation detection and alarm.
40 91 16 for electromagnetic process measurement devices.

26 35 13	Capacitors	
26 35 16	Chokes and Inductors	
26 35 23	Electromagnetic-Interference Filters	
26 35 26	Harmonic Filters	
26 35 33	Power Factor Correction Equipment	
26 35 36	Slip Controllers	
26 35 43	Static-Frequency Converters	
26 35 46	Radio-Frequency-Interference Filters	
26 35 53	Voltage Regulators	

26 36 00 Transfer Switches

Includes: switches transfer from one source of electricity to another.

26 36 13	Manual Transfer Switches	
26 36 23	Automatic Transfer Switches	

26 40 00 Electrical and Cathodic Protection

26 41 00 Facility Lightning Protection

Includes: wiring and equipment for lightning protection.

See: 26 18 19 for medium-voltage lightning arresters.
33 79 00 for site grounding.
33 79 93 for site lightning protection.

26 41 13	Lightning Protection for Structures	
	26 41 13.13 Lightning Protection for Buildings	
26 41 16	Lightning Prevention and Dissipation	
26 41 19	Early Streamer Emission Lightning Protection	
26 41 23	Lightning Protection Surge Arresters and Suppressors	

320

110F47J.EPS

Figure 47 Detailed breakdown of the electrical division (10 of 13).

NUMBER	TITLE	EXPLANATION

26 42 00 Cathodic Protection

Includes: equipment, controls, and installation for cathodic protection of structures and underground metal construction and piping.

See: 40 46 42 for cathodic process corrosion protection.

26 42 13   Passive Cathodic Protection for Underground and Submerged Piping
26 42 16   Passive Cathodic Protection for Underground Storage Tank

26 43 00 Transient Voltage Suppression

Includes: devices to protect against voltage surges on electrical distribution systems.

26 43 13   Transient-Voltage Suppression for Low-Voltage Electrical Power Circuits

26 50 00 Lighting

Includes: luminaries, lighting equipment, ballasts, dimming controls, and lighting accessories. fluorescent, high intensity discharge, incandescent, mercury vapor, neon, and sodium vapor lighting.

Alternate Terms/Abbreviations: HID: high intensity discharge.

See: 10 84 00 for Gas Lighting.
25 36 26 for integrated automation lighting relays.
26 09 23 for lighting controls.
26 20 00 for low-voltage electrical transmission.

26 51 00 Interior Lighting

Includes: lighting for interior locations, except for emergency lighting, lighting in hazardous locations, and special purpose lighting. chandeliers, troffers.

See: 09 54 16 for luminous ceilings.
09 58 00 for integrated ceiling assemblies.
10 14 33 for illuminated panel signage.

26 51 13   Interior Lighting Fixtures, Lamps, And Ballasts

26 52 00 Emergency Lighting

Includes: equipment for exitway lighting and other emergency applications, including emergency battery units, fixtures with integral batter power supplies.

See: 26 53 00 for exit signs.

321

110F47K.EPS

Figure 47   Detailed breakdown of the electrical division (11 of 13).

DIVISION **26**

NUMBER	TITLE	EXPLANATION
26 53 00	**Exit Signs**	Includes: electric exit signs. See: 26 52 00 for emergency lighting.
26 54 00	**Classified Location Lighting**	Includes: lighting for application in areas classified as hazardous. See: 26 55 33 for hazard warning lighting.
26 55 00	**Special Purpose Lighting**	Includes: lighting equipment for specialized applications. Alternate Terms/Abbreviations: healthcare lighting: medical lighting. See: 11 13 26 for loading dock lights. 11 18 00 for security equipment. 11 19 00 for detention equipment. 11 59 00 for exhibit and display equipment. 11 61 00 for theater and stage equipment. 11 70 00 for healthcare equipment. 13 10 00 for swimming pools. 13 12 00 for fountains. 13 14 00 for aquatic park structures. 13 17 00 for tubs and pools. 26 54 00 for classified location lighting. 34 40 00 for transportation signals. 35 13 13 for navigation signals. See Also: 11 61 00 for theatrical lighting.
26 55 23	Outline Lighting	
26 55 29	Underwater Lighting	
26 55 33	Hazard Warning Lighting	
26 55 36	Obstruction Lighting	
26 55 39	Helipad Lighting	See: 34 43 00 Airfield Signaling and Control Equipment
26 55 53	Security Lighting	
26 55 59	Display Lighting	
26 55 61	Theatrical Lighting	
26 55 63	Detention Lighting	
26 55 70	Healthcare Lighting	

322

110F47L.EPS

Figure 47 Detailed breakdown of the electrical division (12 of 13).

NUMBER	TITLE	EXPLANATION

26 56 00   **Exterior Lighting**

Includes: lighting equipment for exterior locations, except for special purpose and signal lighting. airfield general exterior lighting.

Alternate Terms/Abbreviations: athletic lighting: sports lighting.

See: 10 14 33 for illuminated panel signage.
11 13 26 for loading dock lights.
11 68 23 for exterior court athletic equipment.
32 94 00 for planting accessories.
34 41 13 for traffic signals.
34 42 13 for railway signals.
34 43 13 for airfield signals.
34 43 16 for airfield landing equipment.
34 71 00 for roadway construction.
34 72 00 for railway construction.
34 73 00 for airfield construction.
34 75 00 for roadway equipment.

26 56 13	Lighting Poles and Standards
26 56 16	Parking Lighting
26 56 19	Roadway Lighting
26 56 23	Area Lighting
26 56 26	Landscape Lighting
26 56 29	Site Lighting
26 56 33	Walkway Lighting
26 56 36	Flood Lighting
26 56 68	Exterior Athletic Lighting

323

110F47M.EPS

Figure 47   Detailed breakdown of the electrical division (13 of 13).

SUMMARY

In this module, you learned the symbols and conventions used on architectural and engineering drawings. As an electrician, you need to know how to recognize the basic symbols used on electrical drawings and other drawings used in the building construction industry. You should also know where to find the meaning of symbols that you do not immediately recognize. Schedules, diagrams, and specifications often provide detailed information that is not included on the working drawings.

Building projects require detailed specifications. These written specifications are complex and detailed and need a unified format to be easily usable by the trades. The specification format most commonly used is the *MasterFormat*® developed by CSI and CSC. The *MasterFormat*® was updated in 2010 with changes to the division numbering system.

Reading architectural and engineering drawings takes practice and study. Now that you have the basic skills, take the time to master them.

Review Questions

1. A section line on a drawing shows _____.
   a. the north orientation
   b. the location of the section on the plan
   c. where to locate receptacles in that section
   d. the section scale

2. An electrical drafting line with a double arrowhead represents _____.
   a. wiring concealed in the floor
   b. wiring turned down
   c. a branch circuit homerun
   d. wiring concealed in a ceiling or wall

Questions 3 through 9 refer to the seven electrical symbols shown below. In the spaces provided, place the letter corresponding to the correct answer found in the list.

3. ———o——) _____ a. Single Receptacle Outlet

4. ⊖ _____ b. Duplex Receptacle Outlet

5. ◯ PC   ◯ PC _____ c. Triplex Receptacle Outlet

6. ⊕ _____ d. Incandescent Fixture (Surface or Pendant)

7. ⊖ _____ e. Incandescent Fixture with Pull Chain (Surface or Pendant)

Ceiling   Wall

8. ◯   ◯— _____ f. Head Guy

9. ——●—— _____ g. Sidewalk Guy

110RQ01.EPS

---

10. In dimension drawings, the dimensions written on the drawing are _____.

    a. for reference only
    b. on a larger scale
    c. inaccurate
    d. the actual dimensions

11. The architect's scale is designed so that one inch always equals one foot.

    a. True
    b. False

12. All views on a construction drawing are drawn to the same scale.

    a. True
    b. False

13. The *NEC*® specifies one set of electrical drawing symbols that are used in all cases.

    a. True
    b. False

14. Dotted lines used to represent a branch circuit on a drawing mean that the wiring is to be _____.

    a. concealed in the ceiling or wall
    b. run in the floor or ceiling below
    c. exposed
    d. installed in a future building expansion

15. A branch circuit line or drawing that does *not* have slashes is assumed to have two conductors.

    a. True
    b. False

16. To meet general recommendations, a residential branch circuit rated for 2,400VA should have a connected load of no more than _____.

    a. 1,680VA
    b. 1,920VA
    c. 2,040VA
    d. 2,160VA

17. Power-riser diagrams are used to show the _____.

    a. arrangement of electric service equipment
    b. branch circuit layout for power
    c. branch circuit layout for lighting
    d. panelboard schedule

18. The symbols T_1, T_2, and T_3 in a typical motor starter schematic represent _____.

    a. voltage supply lines
    b. auxiliary contacts
    c. motor terminals
    d. line contacts

19. The updated *MasterFormat*® standard _____.

    a. is specified in the *NEC*®
    b. uses a six-digit code for division content
    c. is required by OSHA
    d. allows for fewer subsections

20. The current *MasterFormat*® standard covering communications systems is under _____.

    a. Division 16
    b. Division 27
    c. Division 37
    d. Division 48

Trade Terms Quiz

Fill in the blank with the correct term that you learned from your study of this module.

1. _____ typically include the following information: a site plan, floor plans, elevations of all exterior faces of the building, and large-scale detail drawings.

2. A(n) _____ is an exact copy or reproduction of an original drawing.

3. A simple, single-line diagram used to show electrical equipment and related connections is a(n) _____ diagram.

4. A(n) _____ shows the path of an electrical circuit or system of circuits, along with the circuit components.

5. To convey a substantial amount of detailed information to installation electricians, an engineer will use a(n) _____ drawing.

6. Shown in a separate view, a(n) _____ view is an enlarged, detailed view taken from an area of a drawing.

7. A cutaway drawing that shows the inside of an object or building is a(n) _____ drawing.

8. The sizes or measurements that are printed on a drawing are called _____.

9. The relationship between an object's size in a drawing and the object's actual size is the _____.

10. The height of the front, rear, or sides of a building is shown in a(n) _____ drawing.

11. A building's location on the site is shown in a(n) _____.

12. A drawing that has a top-down view of a building is a(n) _____ plan.

13. A drawing that has a top-down view of a single object is a(n) _____ view.

14. A(n) _____ diagram is a single-line block diagram used to indicate the electric service equipment, service conductors and feeders, and subpanels.

15. Owners, architects, and engineers use _____ to specify material and workmanship requirements.

16. A(n) _____ is a systematic way of presenting equipment lists on a drawing in tabular form.

17. Complicated circuits, such as control circuits, are shown in a(n) _____ diagram.

18. Usually developed by manufacturers, fabricators, or contractors, a(n) _____ drawing shows specific dimensions and other information about a piece of equipment and its installation methods.

Trade Terms

Architectural drawings
Block diagram
Blueprint
Detail drawing
Dimensions

Electrical drawing
Elevation drawing
Floor plan
One-line diagram
Plan view

Power-riser diagram
Scale
Schedule
Schematic diagram
Sectional view

Shop drawing
Site plan
Written specifications

1. A(n) _____ indicates the location of the building on the property.

2. The _____ show the walls and partitions for each floor or level.

3. What are the three main functions of electrical drawings?

_____

_____

_____

_____

_____

_____

_____

4. The title block of an electrical drawing should contain the following ten items:

_____

_____

_____

_____

_____

_____

_____

_____

_____

_____

5. Match the following names to their corresponding electrical drafting lines.

(A) ——————— E ———————

(B) ————————————————

(C) – – – – – – – – – – – –

(D) ————————————○

(E) ————————————●

(F) ————————————▶▶

or

1 2

————————————◣◣

_____ WIRING TURNED UP

_____ BRANCH CIRCUIT HOMERUN TO PANELBOARD

_____ WIRING TURNED DOWN

_____ EXPOSED WIRING

_____ WIRING CONCEALED IN FLOOR

_____ WIRING CONCEALED IN CEILING OR WALL

110WB.EPS

6. What does the letter F stand for in reference to safety switches? _____

7. On a floor plan with a scale of ½" = 1'0", what would be the equivalent distance if you measured 3¾" on the drawing? _____

8. The purpose of a(n) _____ is to identify that part of the project to which the sheet applies.

9. One-line block diagrams are also known as _____.

10. Divisions _____ and _____ of the current CSI specifications cover electrical work.

Wayne Stratton

Associated Builders
and Contractors

How did you choose a career in the electrical field?
Three events in my childhood created the desire to learn the electrical trade. At age six, the farmhouse we lived in was totally destroyed by fire. The cause was electrical. As a young teen, a local electrician had incorrectly wired a heating element and electrocuted several pigs. In 1973, my father hired this electrician to install a motor starter on a grain conveyor. He could not figure it out. I wanted to learn how to do this type of work and do it safely.

Tell us about your apprenticeship experience.
My education is from a technical school. I have attended several manufacturers' training sessions. I had to gain the hands-on experience after learning the trade. My observation of the apprenticeship programs is this: you get hands-on experience while you learn.

What positions have you held in the industry?
I worked as a plant industrial electrician responsible for motor control, DC motors, co-generation, and medium voltage distribution. Later, I began working for an electrical contractor who wanted to expand his business into the industrial field. I worked as a PLC technician designing and installing control systems. In 1987, I began teaching apprenticeship classes.

What would you say is the primary factor in achieving success?
The desire to learn all that I can learn, the ability to think outside the box, and the opportunities to gain a variety of experiences. All this helps me continue to learn and share with trainees.

What does your current job involve?
I teach electrical apprenticeship levels one through four at two different locations in Iowa. My other responsibilities involve task training for electrical licensing, fire alarm, and code updates.

Do you have any advice for someone just entering the trade?
Continue to learn. Completing an apprenticeship program or acquiring an electrician's license is not the end of learning. With code changes every 3 years, there is always more to learn. If you don't understand something, ask! Observe and learn from experienced individuals.

Appendix

Metric Conversion Chart

METRIC CONVERSION CHART

INCHES Fractional	Decimal	METRIC mm	INCHES Fractional	Decimal	METRIC mm	INCHES Fractional	Decimal	METRIC mm
.	0.0039	0.1000	.	0.5512	14.0000	.	1.8898	48.0000
.	0.0079	0.2000	9/16	0.5625	14.2875	.	1.9291	49.0000
.	0.0118	0.3000	.	0.5709	14.5000	.	1.9685	50.0000
1/64	0.0156	0.3969	37/64	0.5781	14.6844	2	2.0000	50.8000
.	0.0157	0.4000	.	0.5906	15.0000	.	2.0079	51.0000
.	0.0197	0.5000	19/32	0.5938	15.0813	.	2.0472	52.0000
.	0.0236	0.6000	39/64	0.6094	15.4781	.	2.0866	53.0000
.	0.0276	0.7000	.	0.6102	15.5000	.	2.1260	54.0000
1/32	0.0313	0.7938	5/8	0.6250	15.8750	.	2.1654	55.0000
.	0.0315	0.8000	.	0.6299	16.0000	.	2.2047	56.0000
.	0.0354	0.9000	41/64	0.6406	16.2719	.	2.2441	57.0000
.	0.0394	1.0000	.	0.6496	16.5000	2 1/4	2.2500	57.1500
.	0.0433	1.1000	21/32	0.6563	16.6688	.	2.2835	58.0000
3/64	0.0469	1.1906	.	0.6693	17.0000	.	2.3228	59.0000
.	0.0472	1.2000	43/64	0.6719	17.0656	.	2.3622	60.0000
.	0.0512	1.3000	11/16	0.6875	17.4625	.	2.4016	61.0000
.	0.0551	1.4000	.	0.6890	17.5000	.	2.4409	62.0000
.	0.0591	1.5000	45/64	0.7031	17.8594	.	2.4803	63.0000
1/16	0.0625	1.5875	.	0.7087	18.0000	2 1/2	2.5000	63.5000
.	0.0630	1.6000	23/32	0.7188	18.2563	.	2.5197	64.0000
.	0.0669	1.7000	.	0.7283	18.5000	.	2.5591	65.0000
.	0.0709	1.8000	47/64	0.7344	18.6531	.	2.5984	66.0000
.	0.0748	1.9000	.	0.7480	19.0000	.	2.6378	67.0000
5/64	0.0781	1.9844	3/4	0.7500	19.0500	.	2.6772	68.0000
.	0.0787	2.0000	49/64	0.7656	19.4469	.	2.7165	69.0000
.	0.0827	2.1000	.	0.7677	19.5000	2 3/4	2.7500	69.8500
.	0.0866	2.2000	25/32	0.7813	19.8438	.	2.7559	70.0000
.	0.0906	2.3000	.	0.7874	20.0000	.	2.7953	71.0000
3/32	0.0938	2.3813	51/64	0.7969	20.2406	.	2.8346	72.0000
.	0.0945	2.4000	.	0.8071	20.5000	.	2.8740	73.0000
.	0.0984	2.5000	13/16	0.8125	20.6375	.	2.9134	74.0000
7/64	0.1094	2.7781	.	0.8268	21.0000	.	2.9528	75.0000
.	0.1181	3.0000	53/64	0.8281	21.0344	.	2.9921	76.0000
1/8	0.1250	3.1750	27/32	0.8438	21.4313	3	3.0000	76.2000
.	0.1378	3.5000	.	0.8465	21.5000	.	3.0315	77.0000
9/64	0.1406	3.5719	55/64	0.8594	21.8281	.	3.0709	78.0000
5/32	0.1563	3.9688	.	0.8661	22.0000	.	3.1102	79.0000
.	0.1575	4.0000	7/8	0.8750	22.2250	.	3.1496	80.0000
11/64	0.1719	4.3656	.	.8858	22.5000	.	3.1890	81.0000
.	0.1772	4.5000	57/64	.89063	22.6219	.	3.2283	82.0000
3/16	0.1875	4.7625	.	.9055	23.0000	.	3.2677	83.0000
.	0.1969	5.0000	29/32	.90625	23.0188	.	3.3071	84.0000
13/64	0.2031	5.1594	59/64	.92188	23.4156	.	3.3465	85.0000
.	0.2165	5.5000	.	.9252	23.5000	.	3.3858	86.0000
7/32	0.2188	5.5563	15/16	.93750	23.8125	.	3.4252	87.0000
15/64	0.2344	5.9531	.	.9449	24.0000	.	3.4646	88.0000
.	0.2362	6.0000	61/64	.95313	24.2094	3 1/2	3.5000	88.9000
1/4	0.2500	6.3500	.	.9646	24.5000	.	3.5039	89.0000
.	0.2559	6.5000	31/32	.96875	24.6063	.	3.5433	90.0000
17/64	0.2656	6.7469	.	.9843	25.0000	.	3.5827	91.0000
.	0.2756	7.0000	63/64	.98438	25.0031	.	3.6220	92.0000
9/32	0.2813	7.1438	1	1.000	25.40	.	3.6614	93.0000
.	0.2953	7.5000	.	1.0039	25.5000	.	3.7008	94.0000
19/64	0.2969	7.5406	.	1.0236	26.0000	.	3.7402	95.0000
5/16	0.3125	7.9375	.	1.0433	26.5000	.	3.7795	96.0000
.	0.3150	8.0000	.	1.0630	27.0000	.	3.8189	97.0000
21/64	0.3281	8.3344	.	1.0827	27.5000	.	3.8583	98.0000
.	0.3346	8.5000	.	1.1024	28.0000	.	3.8976	99.0000
11/32	0.3438	8.7313	.	1.1220	28.5000	.	3.9370	100.0000
.	0.3543	9.0000	.	1.1417	29.0000	4	4.0000	101.6000
23/64	0.3594	9.1281	.	1.1614	29.5000	.	4.3307	110.0000
.	0.3740	9.5000	.	1.1811	30.0000	4 1/2	4.5000	114.3000
3/8	0.3750	9.5250	.	1.2205	31.0000	.	4.7244	120.0000
25/64	0.3906	9.9219	1 1/4	1.2500	31.7500	5	5.0000	127.0000
.	0.3937	10.0000	.	1.2598	32.0000	.	5.1181	130.0000
13/32	0.4063	10.3188	.	1.2992	33.0000	.	5.5118	140.0000
.	0.4134	10.5000	.	1.3386	34.0000	.	5.9055	150.0000
27/64	0.4219	10.7156	.	1.3780	35.0000	6	6.0000	152.4000
.	0.4331	11.0000	.	1.4173	36.0000	.	6.2992	160.0000
7/16	0.4375	11.1125	.	1.4567	37.0000	.	6.6929	170.0000
.	0.4528	11.5000	.	1.4961	38.0000	.	7.0866	180.0000
29/64	0.4531	11.5094	1 1/2	1.5000	38.1000	.	7.4803	190.0000
15/32	0.4688	11.9063	.	1.5354	39.0000	.	7.8740	200.0000
.	0.4724	12.0000	.	1.5748	40.0000	8	8.0000	203.2000
31/64	0.4844	12.3031	.	1.6142	41.0000	.	9.8425	250.0000
.	0.4921	12.5000	.	1.6535	42.0000	10	10.0000	254.0000
1/2	0.5000	12.7000	.	1.6929	43.0000	20	20.0000	508.0000
.	0.5118	13.0000	.	1.7323	44.0000	30	30.0000	762.0000
33/64	0.5156	13.0969	1 3/4	1.7500	44.4500	40	40.0000	1016.000
17/32	0.5313	13.4938	.	1.7717	45.0000	60	60.0000	1524.000
.	0.5315	13.5000	.	1.8110	46.0000	80	80.0000	2032.000
35/64	0.5469	13.8906	.	1.8504	47.0000	100	100.0000	2540.000

TO CONVERT TO MILLIMETERS, MULTIPLY INCHES X 25.4
TO CONVERT TO INCHES, MULTIPLY MILLIMETERS X 0.03937*
*FOR SLIGHTLY GREATER ACCURACY WHEN CONVERTING TO INCHES, DIVIDE MILLIMETERS BY 25.4

110A01.EPS

Trade Terms Introduced in This Module

Architectural drawings: Working drawings consisting of plans, elevations, details, and other information necessary for the construction of a building. Architectural drawings usually include:

- A site (plot) plan indicating the location of the building on the property
- Floor plans showing the walls and partitions for each floor or level
- Elevations of all exterior faces of the building
- Several vertical cross sections to indicate clearly the various floor levels and details of the footings, foundations, walls, floors, ceilings, and roof construction
- Large-scale detail drawings showing such construction details as may be required

Block diagram: A single-line diagram used to show electrical equipment and related connections. See *power-riser diagram.*

Blueprint: An exact copy or reproduction of an original drawing.

Detail drawing: An enlarged, detailed view taken from an area of a drawing and shown in a separate view.

Dimensions: Sizes or measurements printed on a drawing.

Electrical drawing: A means of conveying a large amount of exact, detailed information in an abbreviated language. Consists of lines, symbols, dimensions, and notations to accurately convey an engineer's designs to electricians who install the electrical system on a job.

Elevation drawing: An architectural drawing showing height, but not depth; usually the front, rear, and sides of a building or object.

Floor plan: A drawing of a building as if a horizontal cut were made through a building at about window level, and the top portion removed. The floor plan is what would appear if the remaining structure were viewed from above.

One-line diagram: A drawing that shows, by means of lines and symbols, the path of an electrical circuit or system of circuits along with the various circuit components. Also called a single-line diagram.

Plan view: A drawing made as though the viewer were looking straight down (from above) on an object.

Power-riser diagram: A single-line block diagram used to indicate the electric service equipment, service conductors and feeders, and subpanels. Notes are used on power-riser diagrams to identify the equipment; indicate the size of conduit; show the number, size, and type of conductors; and list related materials. A panelboard schedule is usually included with power-riser diagrams to indicate the exact components (panel type and size), along with fuses, circuit breakers, etc., contained in each panelboard.

Scale: On a drawing, the size relationship between an object's actual size and the size it is drawn. Scale also refers to the measuring tool used to determine this relationship.

Schedule: A systematic method of presenting equipment lists on a drawing in tabular form.

Schematic diagram: A detailed diagram showing complicated circuits, such as control circuits.

Sectional view: A cutaway drawing that shows the inside of an object or building.

Shop drawing: A drawing that is usually developed by manufacturers, fabricators, or contractors to show specific dimensions and other pertinent information concerning a particular piece of equipment and its installation methods.

Site plan: A drawing showing the location of a building or buildings on the building site. Such drawings frequently show topographical lines, electrical and communication lines, water and sewer lines, sidewalks, driveways, and similar information.

Written specifications: A written description of what is required by the owner, architect, and engineer in the way of materials and workmanship. Together with working drawings, the specifications form the basis of the contract requirements for construction.

Additional Resources

This module presents thorough resources for task training. The following resource material is suggested for further study.

National Electrical Code® Handbook, Latest Edition. Quincy, MA: National Fire Protection Association.

Figure Credits

AGC of America, Module opener

John Traister, Figures 6–18, Figures 20–23, Figures 29–31, Figures 33–40, Figures 43–45

Mike Powers, Figure 25, Figure 27

The Groups, Subgroups and Divisions used in this textbook are from *MasterFormat®* 2010, published by The Construction Specifications Institute (CSI) and Construction Specifications Canada (CSC), and are used with permission from CSI. For those interested in a more in-depth explanation of *MasterFormat®* 2010 and its use in the construction industry visit www.csinet.org/masterformat or contact:

> The Construction Specifications Institute
>
> 110 South Union Street, Suite 100
>
> Alexandria, VA 22314
>
> 800-689-2900; 703-684-0300
>
> www.csinet.org, Figure 46, Figure 47

Topaz Publications, Inc., 110SA01

Staedtler USA, 110SA02

Scalex Corporation, 110SA03

CONTREN® LEARNING SERIES — USER UPDATE

NCCER makes every effort to keep its textbooks up-to-date and free of technical errors. We appreciate your help in this process. If you find an error, a typographical mistake, or an inaccuracy in NCCER's Contren® materials, please fill out this form (or a photocopy), or complete the online form at www.nccer.org/olf. Be sure to include the exact module number, page number, a detailed description, and your recommended correction. Your input will be brought to the attention of the Authoring Team. Thank you for your assistance.

Instructors – If you have an idea for improving this textbook, or have found that additional materials were necessary to teach this module effectively, please let us know so that we may present your suggestions to the Authoring Team.

NCCER Product Development and Revision
3600 NW 43rd Street, Building G, Gainesville, FL 32606

Fax: 352-334-0932
Email: curriculum@nccer.org
Online: www.nccer.org/olf

☐ Trainee Guide ☐ AIG ☐ Exam ☐ PowerPoints Other _____

Craft / Level: _____ Copyright Date: _____

Module Number / Title: _____

Section Number(s): _____

Description: _____

_____

_____

_____

Recommended Correction: _____

_____

_____

_____

Your Name: _____

Address: _____

_____

Email: _____ Phone: _____